D1703326

JURE KOTNIK

CONTAINER ARCHITECTURE

THIS BOOK CONTAINS 6441 CONTAINERS.

LINKS

Jure Kotnik
CONTAINER ARCHITECTURE
This book contains 6441 containers.

LINKS International
© Carles Broto i Comerma
Jonqueres, 10, 1-5, Barcelona 08003, Spain
Tel.: +34-93-301-21-99
Fax: +34-93-301-00-21
info@linksbooks.net
www.linksbooks.net

Compiled, edited & written by
Jure Kotnik

Editorial coordination by
Jacobo Krauel

Layout & cover design by
Domen Fras, Aparat.org

Layout consultant
Maja B. Jančič, Aparat.org

English-language editing by
Biljana Božinovski

Front cover photo by
mmw.no > Mike Magler

Cover photos by
Michael Moran, Andre Movsesyan, Illycaffé S.p.A.

Printed in China.

Photo Credits:
© Adolfo Enriquez p.84–85
© AFF architekten archive p.74–77
© Andre Movsesyan p. 160–165, 11 up right, 16 left
© Antje Quiram p. 90–95
© BARK Design collective archive p.224–227
© BOPBAA arquitectura archive p.115, 117 up
© Castor, archive p.232–235
© Christoph Gebler p. 78–83
© Daniel Araño p.242–243
© Danny Bright p.96–99
© Earl Carter p.198–201
© Han Slawik – archive p.68–73
© Illycaffé S.p.A. p.202–205
© Jacob Balt p. 190
© Jones and Partners Architects archive p.250–253, 19 left
© Jongoh Kim Designhouse Inc. p.46–51
© Jure Kotnik p.3–5, 8–9, 10, 12–15, 17, 18 right, 19 left, 21 left, 22–23, 24, 28–29, 118–119, 170, 179, 183 right, 196–197, 254–255
© Keith Dewey p.142–147, p.21 middle
© Lara Swimmer p.130–135
© LOT-EK archive p.214–217, p.246–245
© Luuk Kramer p.184–189
© Maarten Laupman p.100–105
© Marta Cuba p.87, 88, 89 up
© Martyn Wills p.171, 172 left, 173, 176, 177 left down
© Michael Moran p.40–45
© Mikael Olsson p.106–111
© Mike Magler / MMW architects p.30–35
© MMW architects p.210–213
© Morley von Sternberg p.58–61
© Nigel Reid Foster p.124–129
© Normand Rajotte p.11 up middle, 16 right, 154–159
© Oriol Rigat p.238–241
© Paul McCredie p.218–223
© Paul Stankey p.136–141
© Paul Warchol p.120–123
© Peter Aaron (ESTO) p.166–169
© Peter Bennetts p.62–67
© Piotr Bylka p. 177 up, left, down right
© Raul Santalices Agencia4PES p. 87 down right, 89 down
© Rob van Uchelen p.195
© Roderick Coyne p.52–53, 54 down left, down right, 55, 56–57
© Alan Lai p.54 middle
© Ronald Schouten p.193
© Sarah Hewson www.containercity.com p.172 right, 174–175, 177 right up, 178, 180, 182, 183 left
© www.sergebrison.com p.206–209
© Simon Devitt p. 148–153
© T.O.P.office, SABAM, p. 20 left, 36–39, 236–237
© Tempohousing p.191, 194
© Timothy Schenk p.26
© Trimo d.d. p.19 up right
© Vid Brezočnik p.228–231
© Wikipedia,Containerization p.27
© Xavi Padros p.112–114, 116, 117 down
© Zack T. Smith, p.10
In ths book we made all is possible in order to credit and clarify the copyrights with the authors. In case any is missing please contact with the publishing house.

AUTHOR→
JURE
KOTNIK

PUBLISHER→
LINKS
BOOKS

CONTAINER ARCHITECTURE

THIS BOOK CONTAINS 6441 CONTAINERS.

NET
CU.CAP.
28,760
63,400
67,7
2,390

87

28 FOUR
Public Buildings

118 FIVE
Housing Projects

196 SIX
Conceptual & Unbuilt Projects

CHAPTER ONE INTRODUCES CONTAINER ARCHITECTURE.

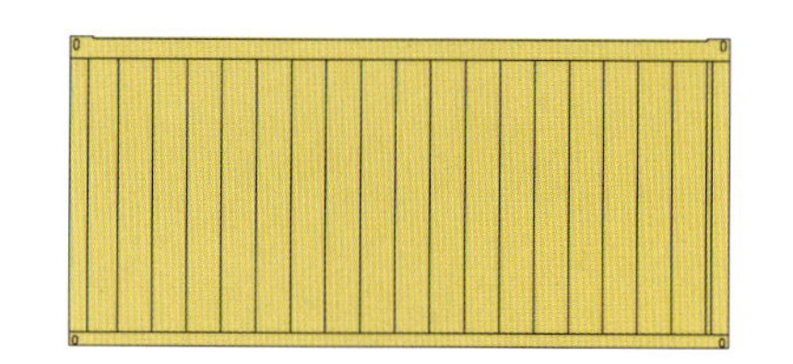

ARCHITECTURE?

A container can easily meet the first two criteria determining a building according to Vitruvius (firmness and usefulness), while the third – aesthetics – is where the architects come in.

CITIZEN OF THE WORLD.

Containers can be found all over the world, which makes container architecture at home all around the globe.

IS IT ARCHITECTURE?

CONTAINER ARCHITECTURE IS A YOUNG BRANCH OF architecture, which has recently witnessed wide media coverage – perhaps disproportionately considering the number of projects that had been realized. The common denominator in container architecture is always the same (the standard ISO container), but the resulting projects are both extremely varied and of high quality, as regards their underlying concepts as well as outlook. The ISO container, a companion of globalization, is a cosmopolitan building block, which makes container architecture at home all around the world.

Container architecture and the ISO container share several characteristics: they are both prefabricated, compact, sturdy, weather-resistant, and can also be mobile or set up only temporarily. Mass production of containers makes this type of architecture affordable, while it has also proved environment-friendly – it recycles and reuses the surplus of shipping containers, which results from the trade imbalance with the Far East. Once its useful features were recognized, this alternative type of construction gradually appealed to both architects and clients, and even became somewhat trendy. Now the number of projects is on the rise, which will help container architecture to become more firmly established in its field, making it less of a sensation and more of a just another legitimate branch of architecture.

Currently virtually all container architecture projects still receive media attention. Because not so many have so far been realized, it was possible to present the most representative ones all in one volume. That this is doubtlessly quality architecture is underlined by the numerous renowned awards these projects have won (one in four projects from the present book have received national or international acclaim).

After an overview of the container system and the historical circumstances that resulted in this type of architecture, the present book gives 45 projects, constructed from as many as 6441 containers. They are grouped into three Chapters according to function, so that there are Housing, Public Buildings and Conceptual Projects, where the last one also includes some as yet Unbuilt projects. The main criteria in selecting projects to be included were: how innovative the concept is, its aesthetics, and how appropriate containers are for a given construction task.

➔TRANSFORMATION.
From a cocoon to the butterfly.

As to the quality of its projects, container architecture is comparable to traditional architecture. It meets all the conditions and has all the qualities of 'perfect' buildings as defined by Vitruvius: firmness and durability (*firmitas*), usefulness (*utilitas*) and beauty (*venustas*). Containers meet the first two principles through their inherent characteristics, while the third – the transformation of a building from a cocoon into a beautiful butterfly – is where architects come in.

➔SPOTLIGHT.
Its fresh approach, the striking difference between the input and output, renowned architects and numerous national and international awards is what makes container architecture attract media attention.

CHAPTER TWO DESCRIBES **CONTAINER ARCHITECTURE.**

Sooner or later architects had to notice the flawless works of the system of giant Lego bricks that travel all around the world and stop in ports, where they are stacked into container "neighborhoods". The associative link with architecture is obvious. Architects have long attempted to find a cost-efficient architectural solution, especially in the area of container housing, that would be easy to transport, modular, prefabricated and mass-produced. But all such potential construction systems failed to succeed. Either the ideas were too revolutionary for the time in which they appeared or they were limited to only small groups of people. Both stood in the way of mass production, which meant an end for the projects because the costs of such buildings were too high. The fact that containers are primarily used in transport and that architects can borrow them as needed is an important advantage. There is namely no need to set up a new system of construction, since this existing one already has all the necessary advantages.

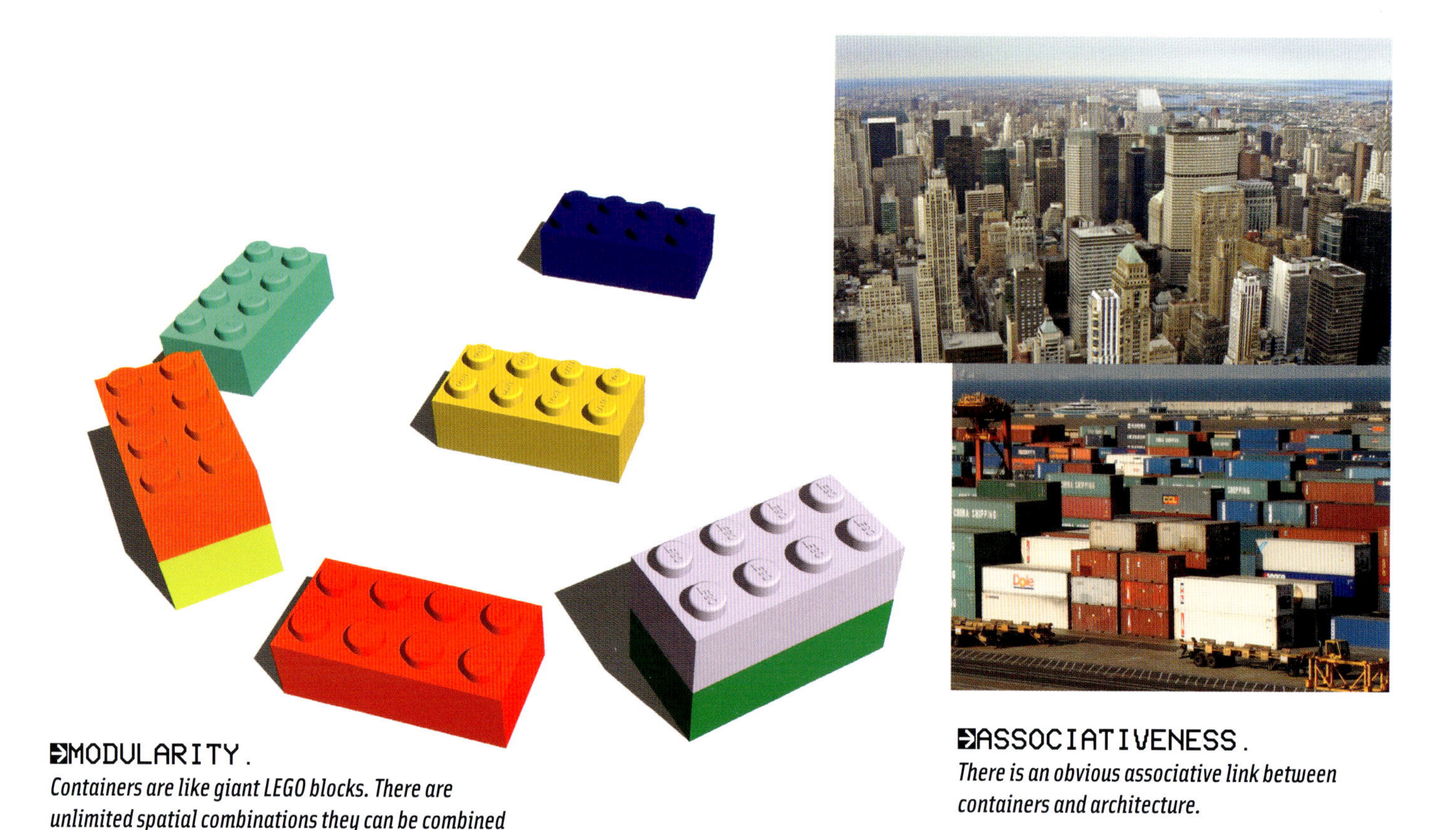

➔MODULARITY.
Containers are like giant LEGO blocks. There are unlimited spatial combinations they can be combined into.

➔ASSOCIATIVENESS.
There is an obvious associative link between containers and architecture.

CHARACTERISTICS

Containers have many characteristics that make them convenient for use in architecture. They are prefabricated, mass-produced, cheap and mobile. Because they are compatible with practically every transport system, they are easily accessible all around the world. They are strong and resistant, while also being durable and stackable. They are modular, recyclable and reusable.

Containers are quite common and relatively cheap – on average a used container costs as little as $1,500, while a new one comes to some $4,000. Using shipping containers for construction purposes can therefore result in lower prices and consequently a more accessible architecture, as well as help solve tight-budget problems. The low costs of this type of construction is also due to the system's modular nature, which enables structures to be disassembled, moved, and then reassembled quickly and with ease. Its modular nature enables a container home to be constructed gradually, depending on the spatial needs of its occupiers, which tend to change during the course of a person's lifetime.

The sturdiness of the containers' outer shell resists any on-site manipulation and withstands the worst of weather conditions – the cold and the heat, as well as salty water, high winds, downpours and other inconveniences. These characteristics are fully shared by container architecture.

At a time when people are becoming ever more environment-conscious, a further advantage of containers is the fact that they are largely recyclable and reusable. Also, if buildings are constructed from containers, the use of other construction materials can be significantly reduced. All this makes container architecture comply with the 3R design concept (reuse, recycle, reduce). Container constructions usually call for no groundwork excavations, which further reduces site impact, and are quick to set up, which means less noise pollution and less waste on the construction site. A smaller container construction can be fully erected within a single day, while larger structures may take up to several days. By way of illustration, the nearly 1,500m² (16,145 sq ft) London Riverside Building (p. 178–183) took no more than a week to complete. Many container projects, most notably the smaller or conceptual ones, strive towards being energy self-sufficient and off-grid by using solar panels and rainwater collectors. Special attention is also paid to interior furnishings, where many environ-

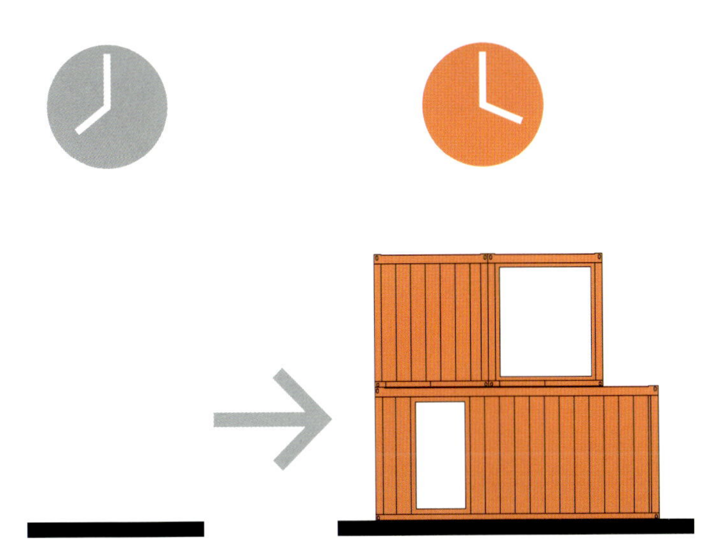

SPEED.

Container buildings, even those of larger proportions, can be assembled in a very short time.

STURDINESS.

A container's outer shell resists any on-site manipulation and withstands the worst of weather conditions (the cold and the heat, as well as salty water, high winds, downpours) and other inconveniences. These characteristics are fully shared by container architecture.

ment-friendly materials are used, such as formaldehyde-free plywood or recycled wood.

Container architecture, nevertheless, also has the other, less appealing side, which architects must handle when converting containers into buildings. Containers need adequate sound-proofing and thermal insulation, and in cases of excessive heat or cold only a sufficiently insulated container is fit to live in. They must therefore be carefully insulated. Some of the materials used for insulating containers are very advanced, such as ceramic-based insulation coating, which comes from aerospace technology and achieves the desired insulation effect with minimum coating. Heat load can also be reduced through a roof or an additional façade layer, which also protects the container from physical force. The façade wrapping is impermeable, which is why good air exchange is urgent either through the windows or mechanically, using proper ventilation systems. Since containers themselves have no openings, the inside is pitch dark. To let in sufficient light, openings for windows and doors are usually cut out. Good planning and execution make these disadvantages easily fixable.

As to construction itself, container architecture is rather straightforward and the building can be delivered on site and welded into place by only a truck and a car lift. As mentioned, converting a container into a home/office includes insulation coating, cutting out window and door openings and doing the floors and inner walls, which is the same as with traditional construction work and therefore attractive as a do-it-yourself project. Stankey Cabin (p. 136–141) and the Zigloo (p. 142–147) are model examples. The construction of the last one was also given extensive on-line coverage, which offered all fans of self-constructed container housing an excellent insight into the erection of a container home.

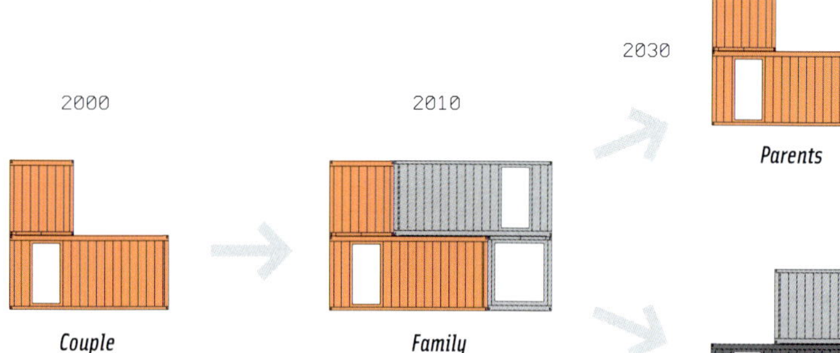

FLEXIBILITY.
Containers are modular in nature. This enables container-buildings' structure to change through time and adapt to spatial needs of their occupiers.

MIXED STRUCTURES.
Container architecture often includes other construction materials aside from containers, such as wood, steel, concrete and glass, which make container-made buildings more like other custom made architecture.

DEVELOPMENT
Container architecture witnessed a kind of bottom-up development, into which architects joined only later. The compact and sturdy transport box, which is weather-, fire-, earthquake-resistant and defies all other types of inconveniences, is spatially suggestive to the extent that people first began putting it to various uses themselves. Especially in Third World countries containers thus spontaneously became shacks, stores and shelters for those who needed them.

At first, containers were closer to architectural/artistic manifestos than architecture proper. They had a powerful concept behind them and highlighted containers' typical characteristics, such as mobility and their cosmopolitan nature, as well as spatial asceticism and minimalistic interior furnishings (Future Shack 1995, p.198–201; Philtex 1994, p.210–213; MDU 2001, p.214–217). Pro-to-users of such dwellings were mostly the so-called urban nomads.

But then container architecture developed further and when the traditional construction materials (wood, steel) joined in the process and made it draw near to the accepted architectural approaches, the circle of potential end-users widened. In the future, container houses are expected to be less experimental and more like the other available custom prefabs on the market (Redondo Beach House, p.160–165), the demand for which is already growing. And all this seems to guarantee that container architecture is here to stay.

A special drive pushing the development of container architecture is trade surpluses. There are redundant containers all around the world (most notably in Europe and North America), due to the imbalance in manufactured goods between the West and the Far East, where the majority of the world's goods are produced. The West imports more than it exports and the imported goods travel in containers. Sending back an empty container costs $900, and it is often more economical to buy a new container in China than to send the used one on the return journey. As a result, the world's ports are piled up with empty containers. In his research paper *Shipping containers as building elements*, J.D. Smith suggests as much as 125,000 abandoned containers to be clogging Great Britain's ports, whereas the number is much higher in the USA, circling around 700,000 (according to Bob Vila). Redundant containers take up space and are po-

INDIVIDUALISM.

In a society stressing the personal style of each individual, containers are perfect for mass-produced customized architecture.

TRENDY.

A house as a product and brand (L. Vuitton ConHouse, 2006).

tentially (as steel waste) a threat to the environment, therefore port operators give them away at low prices. From the standpoint of architectural ethics, it is good that architects did not create the problem, but were merely called upon to find solutions for it when it appeared.

Parallel to the evolution of the cargo container, the system developed other applications as well, such as office and housing containers, which chiefly serve as temporary accommodation – for the army or construction workers, or as emergency and relief housing. These buildings are designed for a market with low standards of aesthetic, which primarily looks for low-cost and functional solutions. This is why containers used for such purposes are very similar all around the world. They are optimized for ground-level use and if joined into larger units without supporting constructions they mainly connect horizontally.

Today's container architecture – both custom and mass-produced – includes: emergency shelters, schools, urban and rural houses, apartment and office buildings, artists' studios, stores, medical clinics, radar stations, shopping malls, markets, exhibition spaces, experimental labs, bathrooms, workshops, press boxes, marines, abstract art, bridges, car salons, bars, restaurants, garages, warehouses, hotels, campuses, kindergartens, galleries, museums, etc.

PREJUDICE.

More and more quality container architecture is successfully fighting prejudice against container constructions.

HOUSING & OFFICE CONTAINERS.

They are useful, cheap and quick to assemble, but lack aesthetics and are therefore not the best representatives of container architecture.

INTO THE FUTURE

Container architecture spans several fields and is bound to develop in different directions. The present popularity of prefab architecture and appearance in Europe and North America of companies specializing in converting used containers and offering container architecture for sale testify to the fact that interest for this type of constructions exists. World trade is expected to result in further imbalances and continue to produce redundant containers in all countries where the service sector is prevalent. The more there are available, the lower the costs of container architecture and consequently the higher demand for it. But though present in large numbers, the available empty containers are not sufficient to turn the tide of architecture and construction business in any significant way by being recycled into buildings; they are more of a supplement to the existing offer of architecture.

From their first editions, McLean's containers gradually improved to what is currently seen as the optimum. Similarly, both the ISO container as such and container architecture in general will continue to gradually evolve and improve, and perhaps even transform into a whole new system. They will mostly retain their current advantages and thus remain prefabricated, quick and easy to set up and transport as well as mass-produced and therefore cheap. But it is also possible that, for instance, the materials will change. As current trends show, another crucial element in the construction business aside from costs is environment-consciousness and energy consumption. It is therefore likely that steel might one day be replaced by another material – something that will take less energy to manufacture, while still being recyclable and reusable just like the ISO container.

There are several possible scenarios which way container architecture could go, all taking account of the modern trends in society. Parallel to the latest advances in the consumer-oriented automotive industry, home-shopping a house should become as simple as selecting the components you want your car to come with. People/clients will browse the Internet to find the rooms they need and online architects will give advice on spatial solutions and interior design. Interior designers and furniture manufacturers will also take part by offering their products. Home-shoppers will direct the entire process themselves and thus get tailor-made homes meeting their specific needs.

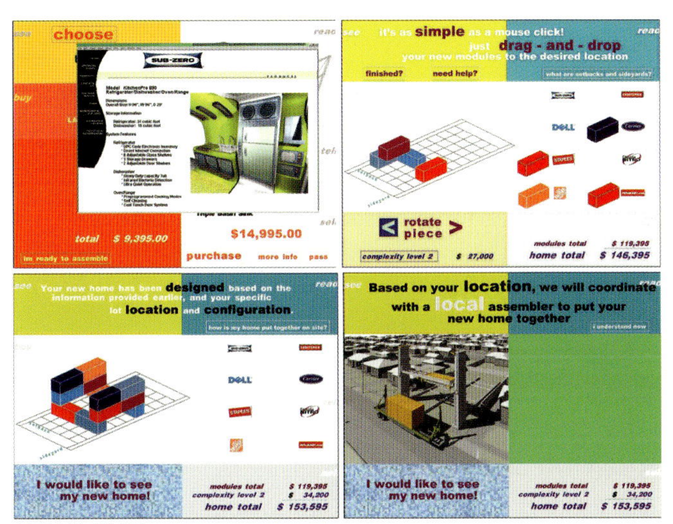

ONLINE.

Buying a contemporary home should become simple enough; you will be able to click your way through the online home-shopping process from the comfort of your favorite armchair.

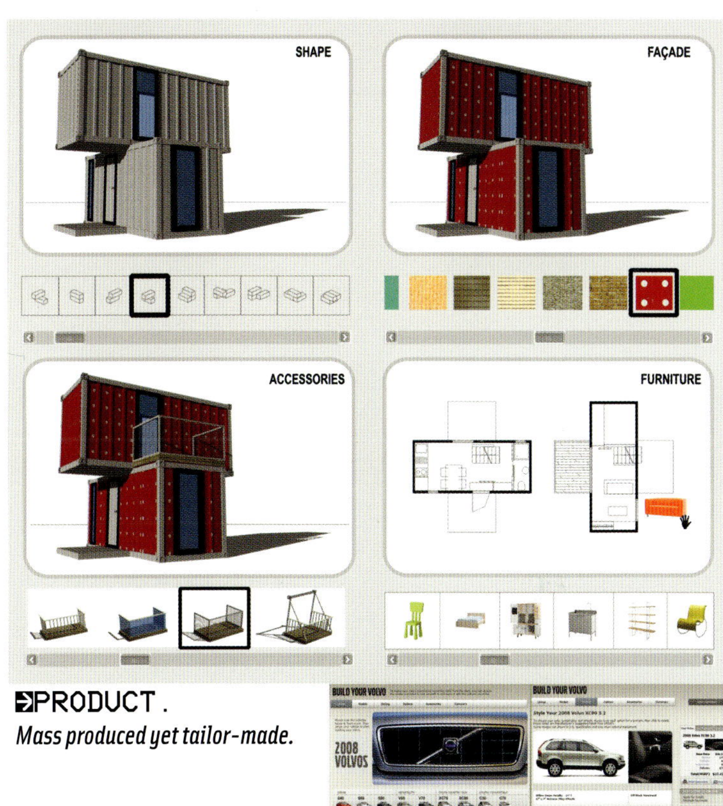

PRODUCT.

Mass produced yet tailor-made.

Ready-made container compounds will be available, for people to buy independently and either supplement their existing home with one or two, or combine several to create their tailor-made home from scratch. Instead of choosing their bathtub, washbasin, shower cabin and bathroom tiles, etc separately, people will order the entire container-fitted ready-made bathroom a-la-carte; buying a container-wellness studio, container-bedroom, container-kitchen, etc will be available at the click of the computer mouse (PRO/CON Package Home Tower, p.250–253). Container-saunas, for instance, already exist (Saunabox, p.232–235).

Despite their numerous advantages, the reality is such that many people still have deep-seated prejudice against containers. Their breakthrough into the mainstream could thus largely be aided by famous designers and architects. Owning a container home by a renowned architect would be similar to owning a Louis Vuitton purse – it would become a status symbol and stir popular demand. The Swedish clothing giant H&M were thinking along similar lines when they invited Stella McCartney, Karl Lagerfeld, Roberto Cavalli, etc to design clothes for them – the point was to create quality clothes at affordable prices targeted at the wide public. Here the result would be quality homes of a trendy design at attractive prices, targeted at the widest possible home-shopping market.

ART.

The container architecture movement started with artistic and architectural manifestos. (Luc Deleu: Speybank (1999); Middelheim Museum, Antwerpen, Belgium)

ENVIRONMENT-FRIENDLY.

Containers are highly recyclable and reusable. The use of containers in construction reduces the use of other materials and has low site impact (less waste and noise pollution on the construction site as well easy dismantling).

PLACE IN SOCIETY

Container architecture has drawn attention mainly from people close to the world of design, fashion, architecture, and those who can appreciate fresh and trendy concepts. Because fresh and trendy is what containers definitely are, since this is what the media has made them to be. A container has everything such people could want; it is mobile, autonomous, flexible, environment-conscious and has the charisma of a true cosmopolitan. Before becoming a part of a house, a container probably sailed the oceans and visited the world's largest ports, not to mention what it carried in its interior. Reena Jana puts it vividly in BussinessWeek, where she points out: "How many people can say the previous occupiers of their home were 20,000 toy dolls, 6,000 pairs of sneakers, or 500 computer monitors? Or that their house had been to China and back?"

Unfortunately, container architecture is still largely associated with emergency housing and monotonous standard housing containers that are poorly illuminated and furnished with materials of lower quality (if compared to standard architecture). The fact that containers are limited in width to just below 2.5m (8 ft) does not help them win popular acclaim either. It is precisely this monotony and absence of aesthetics in such standard container buildings that stand in the way of widely introducing container architecture into the postmodern society, which rests on individualism and personal style. Luckily, architects of custom-made container buildings managed to avoid these traps; the projects given in the book present a variety of options containers can be put to in architecture, and are very different from each other. And each good new product helps container architecture to become more popular with the people and closer to being accepted as a valid construction technique.

Fortunately, containers are cheap to build with. And the language of money is one that people understand well. Sky high prices of real estate in the western world have stimulated the search for and development of alternative construction solutions, and one such attempt is also container architecture. In the capitalist society of today, what is important is the ratio between people's income and the costs of their homes. If their homes are too expensive, they cannot spend for instance on themselves to develop their personal style, which is

PROCESS.

Converting a container into a building includes insulation coating, cutting out window and door openings to ensure proper illumination of the interior, and doing the floors and inner walls.

seen as important in a consumer-driven society. But with budget-friendly containers, what people save up from having a container home as opposed to a traditional house they can use to expand their living space or invest more into interior design. With public buildings, the difference in price allows for the use of quality details, such as for instance the illuminated roof covering Hamburg's Cruise Center (p.78–83) or the decorated meshed tent-like structure invigorating the Fawood Children's Centre (p.52–57).

SURPLUS.

Due to trade imbalances with the Far East, US and European ports abound in redundant containers that can be used as cheap construction material.

CHAPTER
THREE
PRESENTS
THE
ISO
CONTAINER.

The range of container housing is wide and spans from small-size retreat cabins to container villas and large apartment buildings. Its history has seen three main stages. First there were attempts how to fit an entire apartment into a single steel box. Such apartments were designed mainly for the so-called urban nomads – the "side-products" of the modern society, however inhumane this may sound. Then there came smaller holiday houses created and owned mostly by architects and designers who thus paid tribute to this simply clever idea of living in a container. These are the people with a refined taste for what is trendy and containers are currently very "in". Container housing has become more widely acceptable for the public in the third stage, when containers are being combined with other construction materials, the result of which is houses that are quite similar to other custom housing architecture and containers only spice them up a bit.

Small-size container houses are still largely client-oriented and custom-made. The construction of container-made apartment buildings, however, is driven mainly by their practical value and economic efficiency, both for investors and users. The largest of container-made apartment buildings comprise as many as 1,000 units. The modular monotony of such a vast number of identical elements is broken down by diverse facades and installation patterns.

OUTSIDE.

Steel framework, thin-plate outer shell and roof, container doors, markings.

INSIDE.

Most products we use daily once lived inside a container.

ISO CONTAINER

The ISO container is the steel box that paved the way for the globalized society of today and that resulted not only in container architecture but also had an indirect impact on architecture as a whole. The noun *container* derives from the verb *to contain* and accurately denotes the containers' primary function – to hold (and transport) goods. Only a minor segment of the world's containers are used for work/live spaces and for other purposes.

In general, the ISO container consists of a steel framework, thin plate outer shell and roof, and mainly wooden floors on a steel base. Over the years containers have been refined to eliminate any structural/material redundancy. The result is a ready-made space, which a few touches (insulation and proper lighting) can transform into interesting architectural material. The elementary container construction is twice as solid as any building code requires, which makes it appropriate as a building block even without modifications.

Containers are weather-proof and defy hurricanes, floods and numerous other inconveniences (due to their low weight, they are also earthquake-resistant), as well as being fire-proof thanks to the special coating on the façade. Being a closed steel box, they likewise prove a good solution to termite or rodent problems.

THE FATHER.
Malcom Mclean introduced the world to shipping containers as we know them today.

HISTORY.
Several transport systems were used for carrying goods throughout history. The ISO container has prevailed.

HISTORY

The first predecessors of containers appeared as early as the first civilizations. People used various vessels and holders to store and transport goods. The largest push came with the industrial revolution and the appearance of rail transport.

The father of the modern container as we know it today is Malcom McLean, who is often referred to as the father of "conterization". He was an American truck driver and entrepreneur who thought of a way to reduce the costs of and shorten the time-consuming and labor intensive task of unloading the individual contents of a truck onto a ship. His idea was that the entire truck trailer itself could be loaded onto and off of a ship, which would significantly speed things up. In 1950s he took out a patent for a metal shipping container with reinforced corner posts that could be craned off a truck chassis and had integral strength for stacking. This revolutionized the entire transport branch. There were several container systems in use back then, but gradually McLean's containers prevailed. They improved over the years and evolved into a unified system governed by international ISO standards.

Not everyone liked the idea at first, of course; dockers feared they would lose their jobs and port operators had to invest into cranes and other expensive equipment for loading containers onto and off of ships. But the advantages were too promising to be overlooked. McLean's container was a guarantee that the shipment would arrive to the end-destination undamaged. Transport became safer because containers were closed boxes and the cargo did not show, which was less tempting for robbers. Goods also reached their destination sooner this way, because the loading process was now much shorter; one of the significant contributions of containers is precisely their shortening of the transport chain.

Shipping containers played a crucial part in facilitating global trade. By keeping transport and communication costs at bay as well as by widening the world's port distribution channels, they remain the ones that maintain it in full swing.

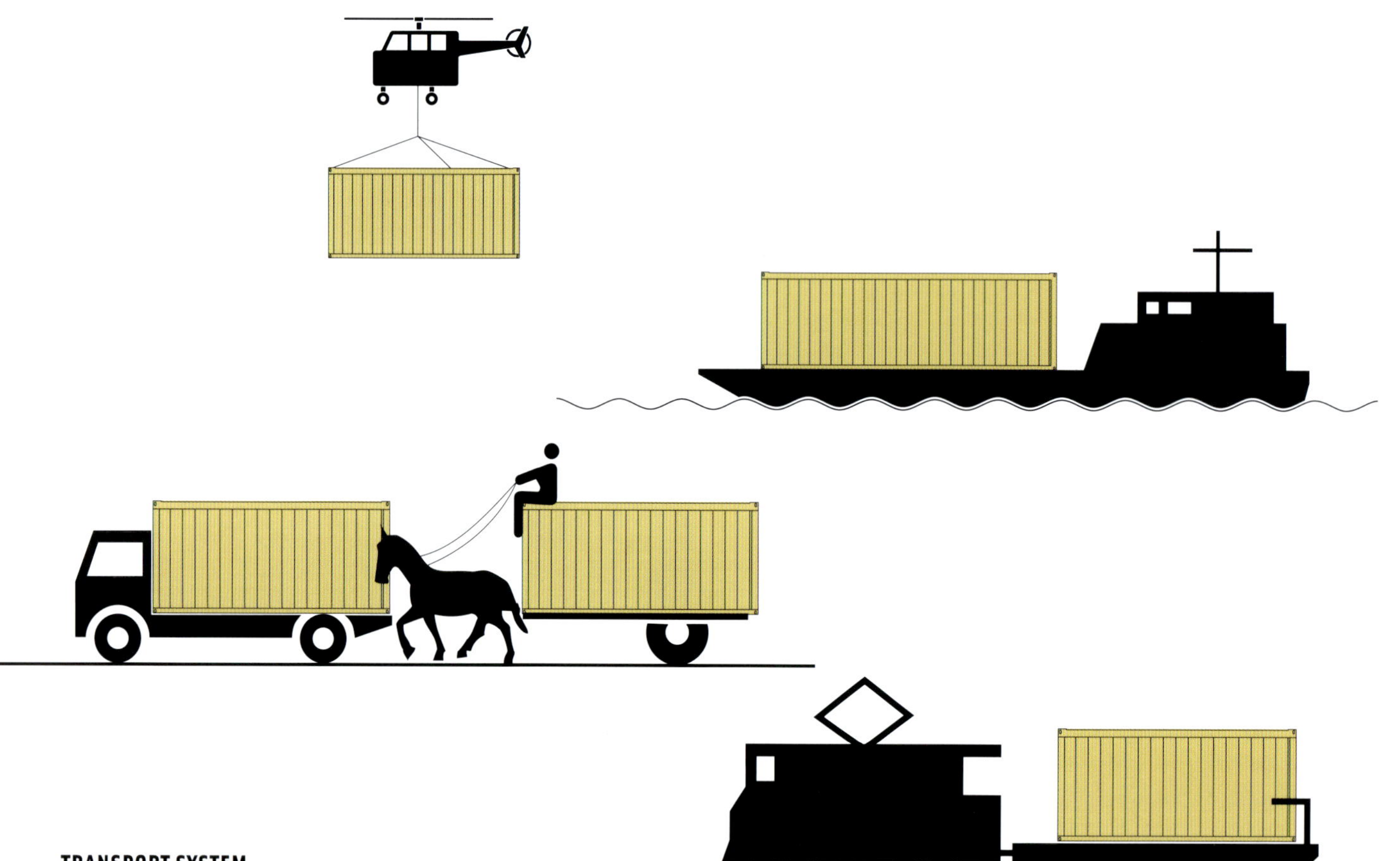

TRANSPORT SYSTEM

According to estimates, some 18 million containers take over 200 million trips every year. The majority of today's bulk cargo is transported in containers and there is hardly a product that has never seen the inside of one.

Transport containers come mainly in two lengths – either ISO 20' (6.05m) or ISO 40' (12.10m) – although others of different lengths and various heights are also in use. They all share the same width of 8 ft (2.43m). There have been tendencies to extend and widen containers beyond the adopted ISO regulations, mainly in the USA, where roads allow for the so-called "jumbo" containers. These are 8 ft 6 in wide (2.6m) and 9 ft 6 in high (2.90m) and extend from 45 ft (13.72m) up to 53 ft (16.15m) in length.

➔MOBILITY.

Containers are well-suited to all means of transportation.

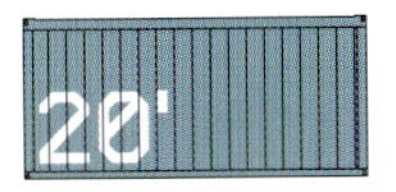

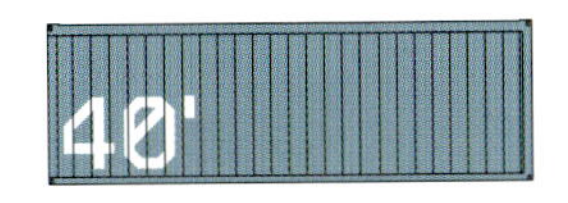

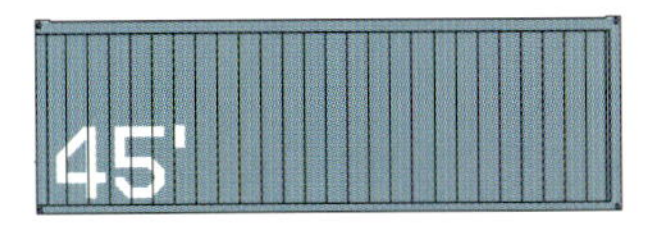

		20′ container		*40′ container*		*45′ high-cube container*	
		imperial	metric	imperial	metric	imperial	metric
external dimensions	length	19′ 10″	6.058 m	40′ 0″	12.192 m	45′ 0″	13.716 m
	width	8′ 0″	2.438 m	8′ 0″	2.438 m	8′ 0″	2.438 m
	height	8′ 6″	2.591 m	8′ 6″	2.591 m	9′ 6″	2.896 m
interior dimensions	length	18′ 10 5/16″	5.758 m	39′ 5 45/64″	12.032 m	44′ 4″	13.556 m
	width	7′ 8 19/32″	2.352 m	7′ 8 19/32″	2.352 m	7′ 8 19/32″	2.352 m
volume		1,169 ft^3	33.1 m^3	2,385 ft^3	67.5 m^3	3,040 ft^3	86.1 m^3
maximum gross mass		52,910 lb	24,000 kg	67,200 lb	30,480 kg	67,200 lb	30,480 kg
net load		48,060 lb	21,800 kg	58,820 lb	26,680 kg	56,620 lb	25,680 kg
area	exterior	158,972 sq ft	14,769 m^2	339.041 sq ft	31,498 m^2	359,934 sq ft	33,439 m^2
	interior	145,764 sq ft	13,542 m^2	304.607 sq ft	28,299 m^2	343.185 sq ft	31,883 m^2

TABLE.

The table shows the weights and dimensions of the three most common types of containers worldwide. The weights and dimensions quoted below are averages, different manufacture series of the same type of container may vary slightly in actual size and weight. The amount of cargo transported in containers is measured in TEUs, which stands for Twenty-Foot Equivalent Units. The largest ships can carry some 14,000 TEU (14,000 20′ containers).

DIVERSITY.

Different colors, lengths, logos and markings distinguish one container from the next and create colorful mosaics that embellish the world's ports.

CHAPTER
FOUR
SHOWS
**SIXTEEN
PUBLIC
BUILDINGS.**

Public buildings have had special significance throughout history and as such have differed from other buildings due to their size, appearance or materials used. From this viewpoint, containers may seem inappropriate for such projects, but in the present Chapter architects make us see this is not true. Well thought-out architecture and carefully chosen details breed buildings that rest on sound conceptual foundations and boast a high level of aesthetics. The interiors of these buildings not only serve their purpose, but are also lavish in space and some even seem like cathedrals for the 21st century. This goes to show that it is not *what* you use, but *how* you use it. Many container-made public buildings answer to the need for temporary spatial solutions, serving their purpose but for a few years time or even less. Containers are also becoming popular in event architecture, where their added value is the fact that such structures – no matter what the proportions – can be assembled and dissembled very quickly. What is more, deciding on a container-constructed public building can save up for an expensive detail or two, which gives the building its 'soul', so to say.

Projects in this Chapter clearly demonstrate that an innovative approach combined with a feel for architecture results in high-quality public buildings, for some of which it is often a pity they are not there to stay.

At the end of a pier in Oslo's Tjuvholmen harbor rises a container gallery that goes by the name GAD. The client's request for a structure of a mobile character was realized though the use of grey steel and glass containers. Due to a special infrastructure facilitating easy assembling and dissembling, the three-level construction can easily be transported and set up at another location within a matter of days.

Its first placement on a pier overlooking the Oslo fjord was not incidental. This is the place where many of Oslo's most beautiful ships were built in centuries past, and to pay tribute to traditional craftwork, the semi-temporal gallery includes shapes that are reminiscent of the ship's, such as the circular windows placed opposite each other on the first floor. Industrial ladders/stairs and containers provide another link with the shipbuilding and transport industry.

The gallery consists of 10 ordinary steel containers composed in the manner of the "Jenga" game. The largest gallery space is on the ground floor, which consist of five 20' containers. These support the first floor, comprising three 40' containers taking the shape of the letter U. This is where the gallery has its entrance, accessible through an exterior staircase. The first floor hosts the reception desk, service room, storage and a small gallery space that includes a staircase leading further up to the second floor gallery, composed of the remaining two containers. The interior is insulated and covered with sheets of plywood and drywall, with the walls being decorated in artworks. All the rooms have plenty of magical northern light, which is invited into the gallery through huge safety glasses at the end of each container, which create an open, breezy feeling, and open up a view of the sky and the sea.

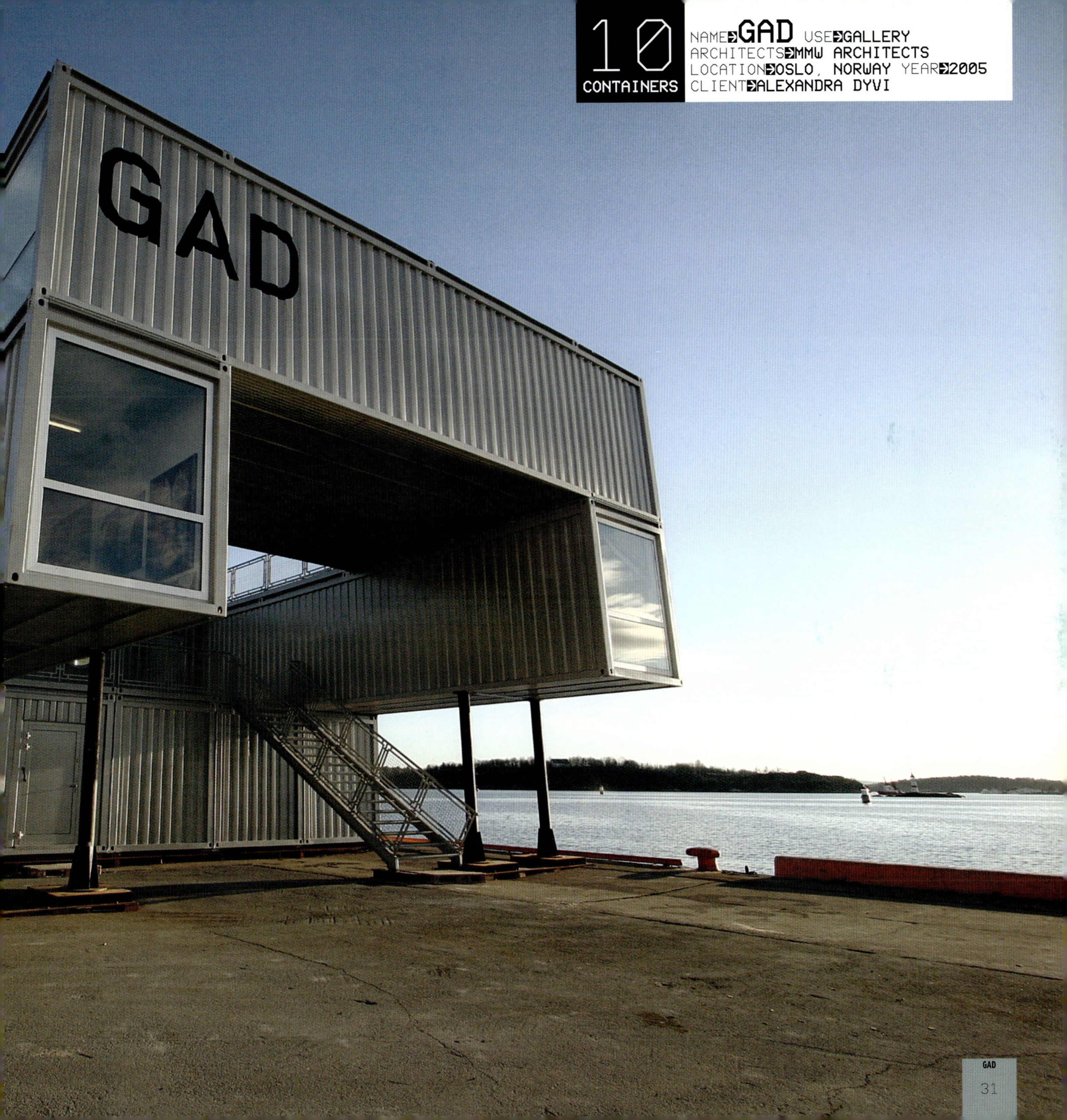

10 CONTAINERS

NAME GAD USE GALLERY
ARCHITECTS MMW ARCHITECTS
LOCATION OSLO, NORWAY YEAR 2005
CLIENT ALEXANDRA DYVI

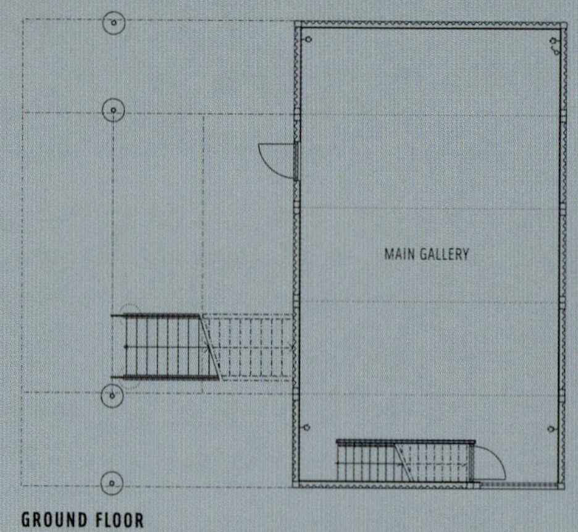

GROUND FLOOR

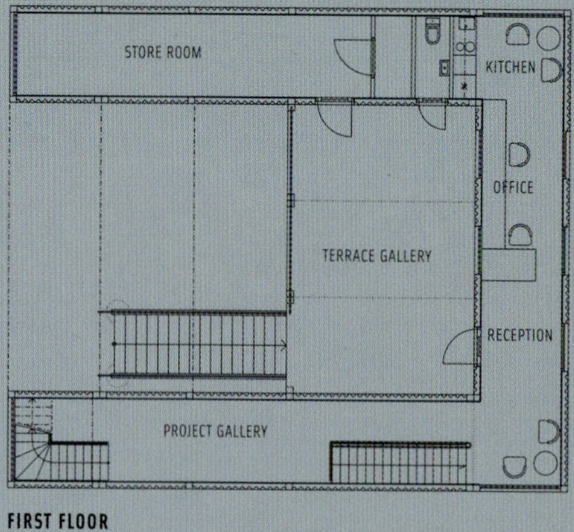

FIRST FLOOR

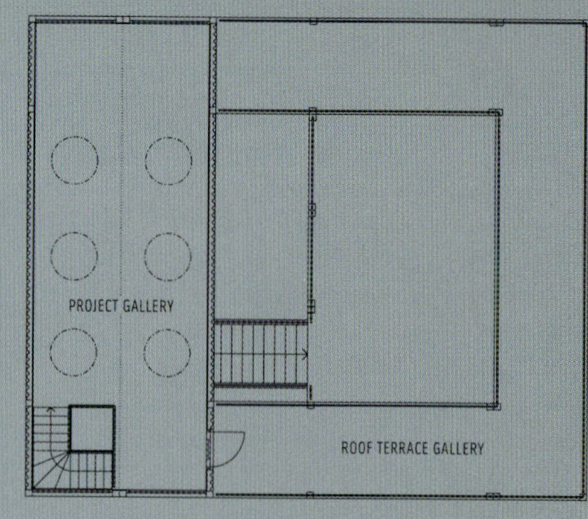

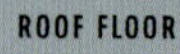

GAD
GAD

Containers have continuously manifested their adequacy as venues for exhibitions and art galleries, since they are quick to assemble and remove from site, as well as easy to transport. Sometimes the fact that a gallery is really a shipping container is a statement in itself, a symbolic representation of the consumer society and its omnipresent problem of redundant cargo containers. Artists are often tempted to reuse such containers for a new purpose, creating fascinating examples of industrial architecture, various installations or simply monuments to one of the world's most efficient systems. The Orbino gallery was first built for the exhibition "Van IJ tot Zee", on top of a dumping ground in Nauerna, near Amsterdam, in 2002.

Orbino consists of five containers, painted red. Two upright containers support a vertical gallery of three, creating a completely asymmetrical structure that visually appears impossible. But the composition is stable, since the forces of the pavilion pass over the anchors down into the foundations and thus defy gravity, wind load and mobile load. The entrance is on the second floor at the back, at the top of a steel staircase, with the front side being glazed and offering a fantastic view of the surroundings.

Orbino proved a successful mode of composition, having been re-enacted twice afterwards – in 2004, the open-air sculpture museum Middelheim, Antwerpen, made Orbino part of their permanent collection, while in 2006 it was rebuilt in Alkmaar where it serves as an "art laboratory", in which artists can realize temporary projects.

NAME KUNSTLAB ORBINO
USE PRESENTATION GALLERY ARCHITECTS LUC DELEU - T.O.P. OFFICE
LOCATION NAVERNA, AMSTERDAM, THE NETHERLANDS YEAR 2002
CLIENT STICHTING KUNST EN CULTUUR NOORD-HOLLAND OF HAARLEM

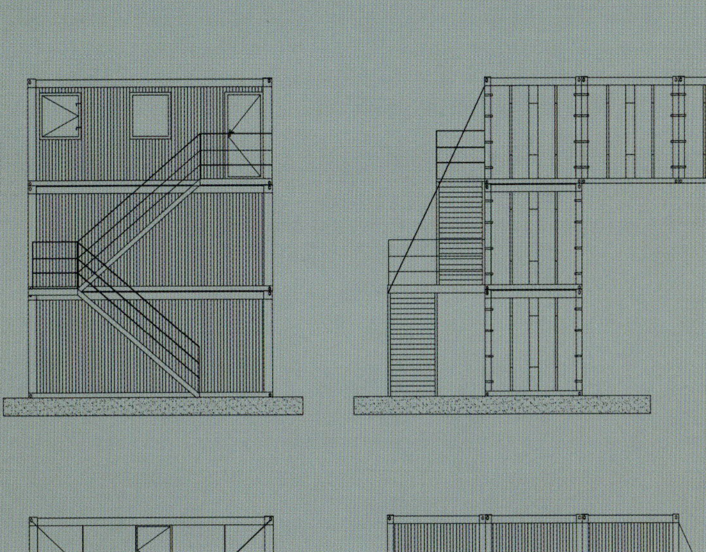

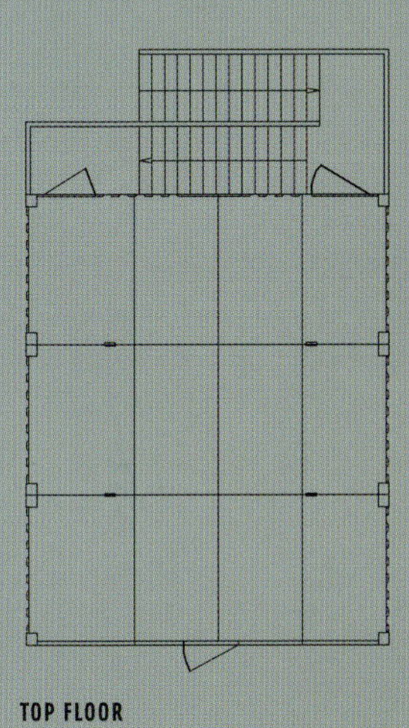

TOP FLOOR

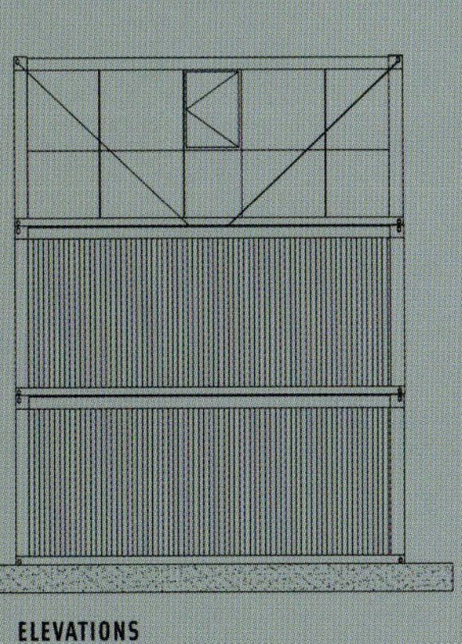

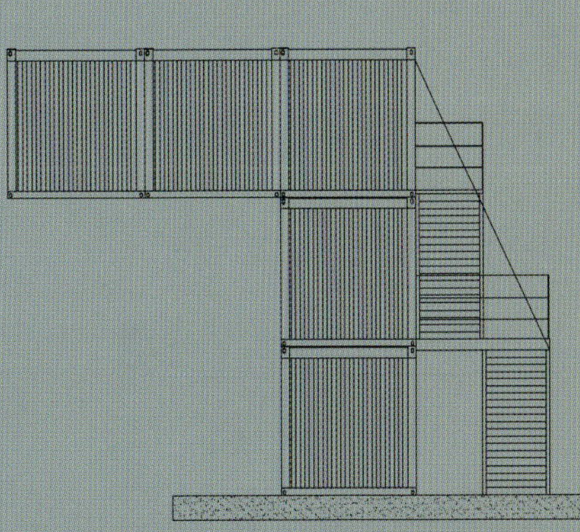

ELEVATIONS

Gregory Colbert, author of the ongoing project Ashes and Snow, approached architect Shigeru Ban about creating an environment-friendly, sustainable, innovative construction to house its show and accompany it in its migratory journey around the world. Colbert wanted a cathedral for the 21st century and he got one – the Nomadic Museum.

This building's journey started on the abandoned Pier 54 of New York's Hudson River. It is composed of 148 borrowed and reusable cargo containers, stacked into a four-level self-supporting grid. The opening spaces between the containers are secured with a diagonal tent-like fabric, creating the typical checkerboard pattern over the entire facade. The interior construction consists of triangular trusses of paper tubes resting on a colonnade of 11m-high (36 ft) paper tube columns. A wooden walkway runs the length of the colonnade and is bordered on both sides by bays full of white-washed river stones, surrounding each column. Above these bays, large unframed photo artworks printed on hand-made Japanese paper are hung on thin cables, installed between the paper columns. This establishes a visual boundary between the physical space of the public walkway and the mystical domain of the images. The fact that shipping containers were used to make this monumental structure does not diminish the awe-inspiring experience of its interior. The atmosphere created by the paper tubes colonnade, stacks of steel containers and carefully placed lights give the visitor the impression of walking into a cathedral – a cathedral honoring art.

The reason why containers were chosen as chief construction elements is the fact that they are readily available in every place the museum travels to. Rather than having to ship the entire exhibit, only 37 containers have to be transported each time, to pack the fabric and materials for the structure, with the remaining ones being borrowed at each new location. The Museum was so far erected in New York City, Santa Monica in California and Tokyo, each time in a slightly different cast of containers, always adapting somewhat to its current environment.

148 CONTAINERS

NAME: NOMADIC MUSEUM
USE: MUSEUM ARCHITECTS: SHIGERU BAN ARCHITECTS
LOCATION: NEW YORK, NJ, AND OTHER YEAR: 2005
CLIENT: PRIVATE

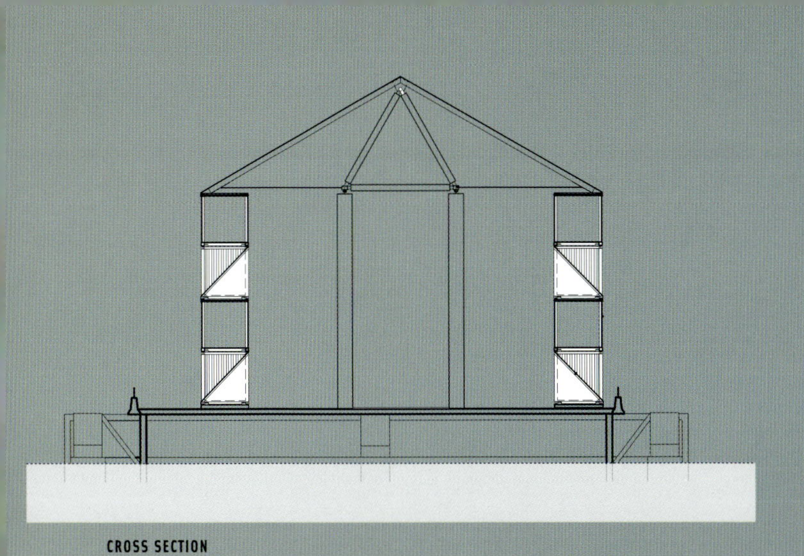

CROSS SECTION

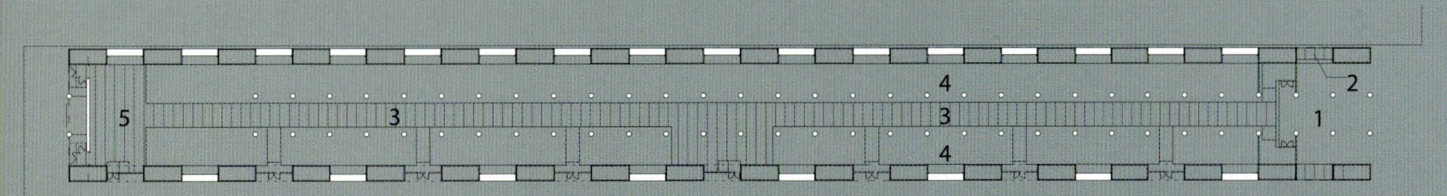

FLOOR PLAN

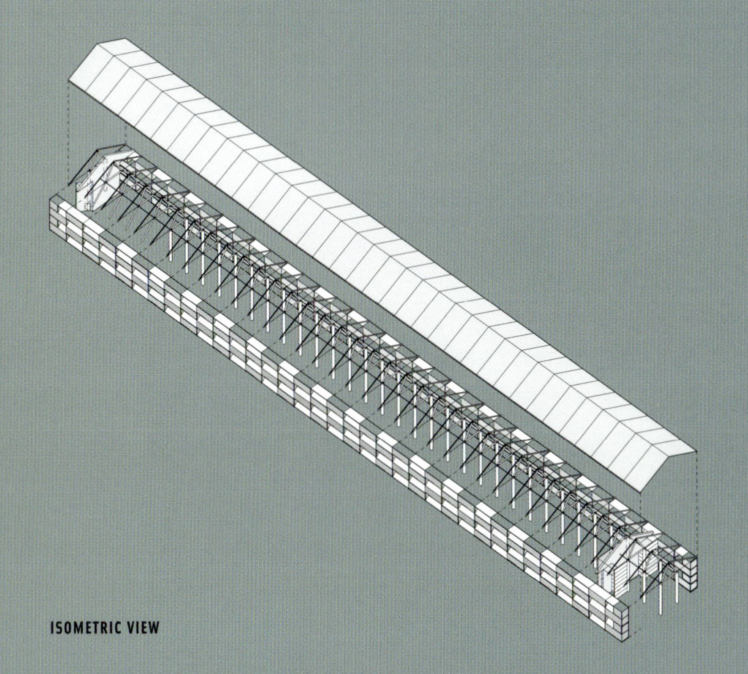

ISOMETRIC VIEW

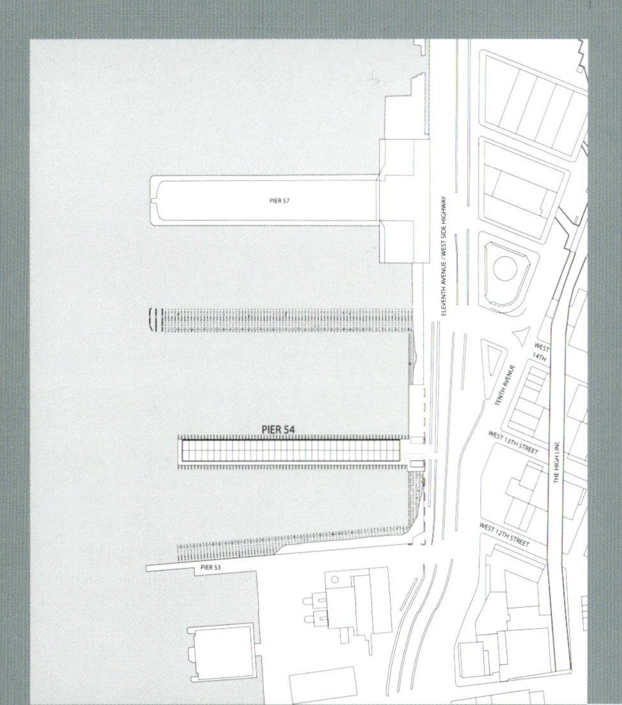

SITE PLAN

TIPHOOK
TRITON

Located in the pleasant landscape of the Olympic Park in Seoul, the Papertainer Museum was commissioned by the Korean design & cultural contents media corporation Designhouse Inc. to commemorate its 30th anniversary of establishment. It is constructed from paper tubes and containers, and thus symbolically represents a meeting place of the past and the present. Paper, the civilization's messenger, and containers, which signify trade, the modern world and the globalized reality of today, together symbolize the world's culture and art.

The Papertainer consists of 166 cargo containers and 373 paper tubes. The paper tubes reach 75cm (2.5 ft) in diameter and 10m (33 ft) in height, and form a columnar system typical of traditional Korean architecture. The ground plan takes the shape of the letter D, where the semi-circle is made of paper. The orthogonal part of the museum is a container hall constructed from four levels of containers and topped off with yet a fifth layer that, together with tent-like fabric, makes up the roof.

The duality of the construction materials is further reflected in the division of the museum into two main exhibitions. The first gallery is the Paper Gallery, where the works appear on paper hanging from the ceiling. It exhibits photographs and paintings of famous female figures from Korean folklore and history, as recently recreated by 30 renowned artists, and is titled "Spotlight on 30 Women". The second gallery is the Container Gallery. It hosts the 'Spotlight on 30 Brands', displaying the brand images of 30 corporations, as re-interpreted and made into pieces of art by contemporary Korean designers. The building is all about binary pairs: two different exhibitions represent two different worlds – the old one and the new one – in two different environments and through two different core materials.

Like its peer project, Shigeru Ban's Nomadic Museum, Papertainer Museum is planned to tour other locations, across Korea as well as abroad.

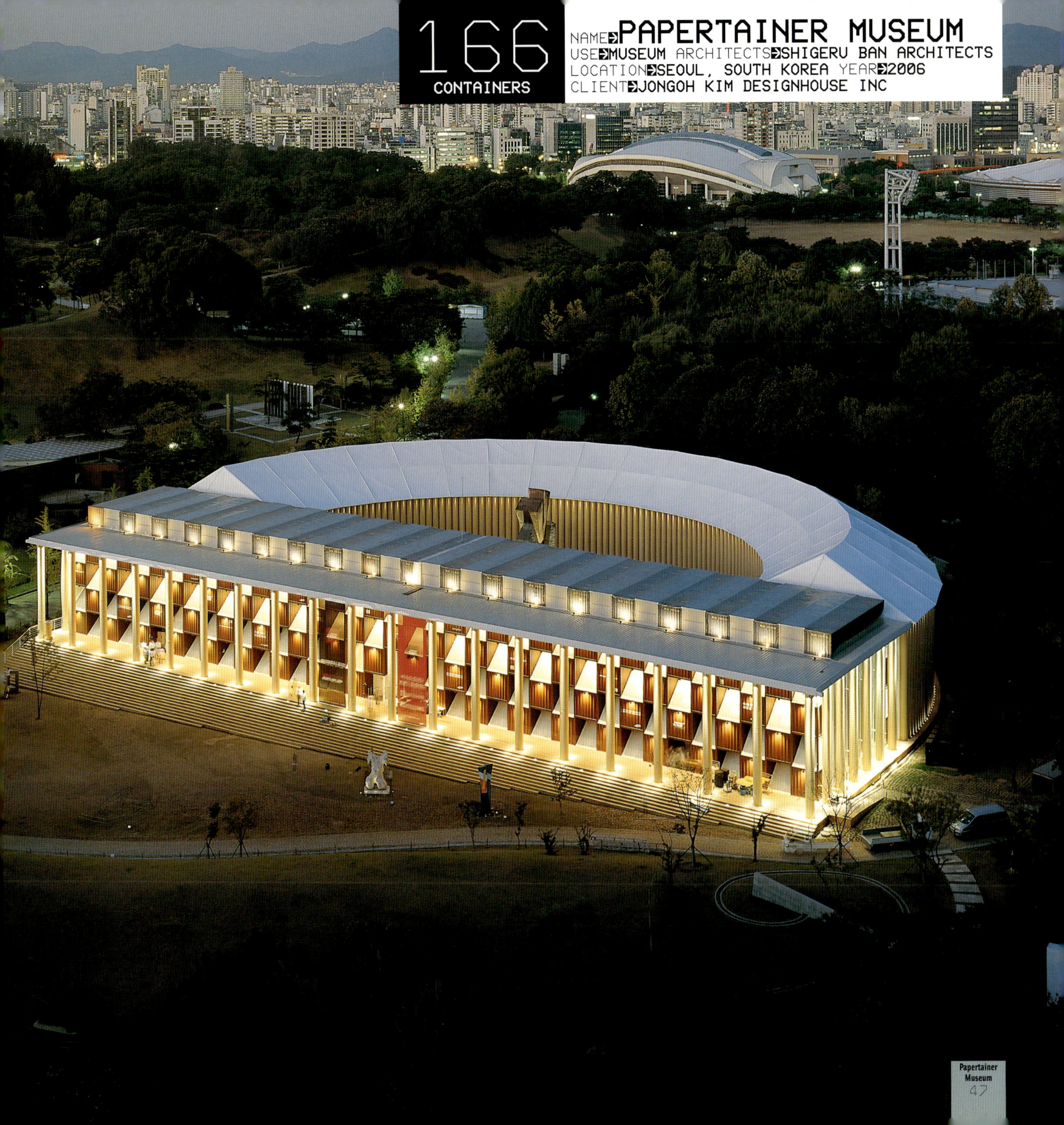

166 CONTAINERS

NAME ➔ PAPERTAINER MUSEUM
USE ➔ MUSEUM ARCHITECTS ➔ SHIGERU BAN ARCHITECTS
LOCATION ➔ SEOUL, SOUTH KOREA YEAR ➔ 2006
CLIENT ➔ JONGOH KIM DESIGNHOUSE INC

SPOTLIGHT 30 WOMEN
SPOTLIGHT 30 BRANDS
PAPER TAINER
20060915-2007-

NORTHEAST ELEVATION

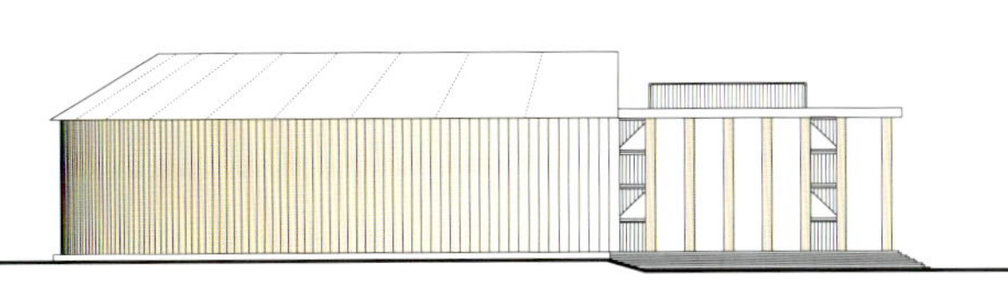

NORTHWEST ELEVATION

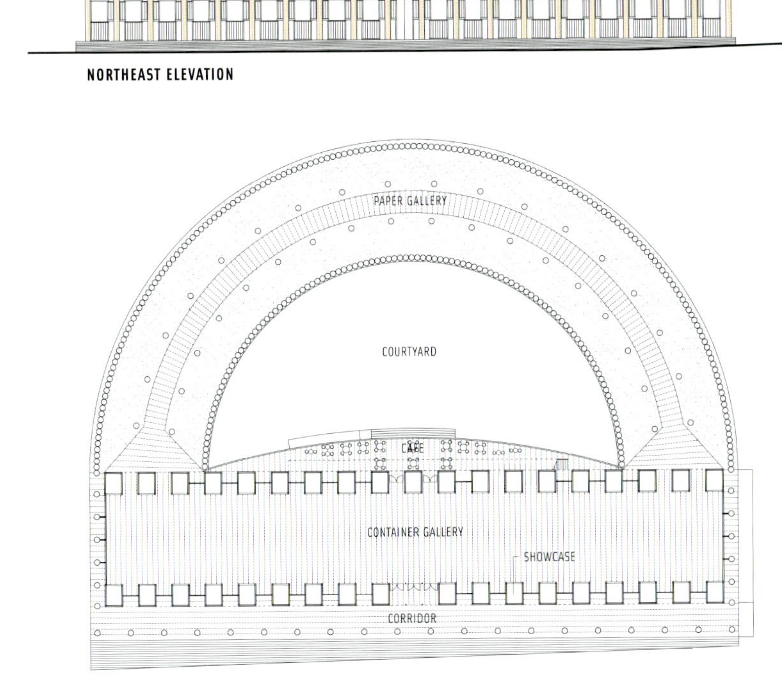

FLOOR PLAN

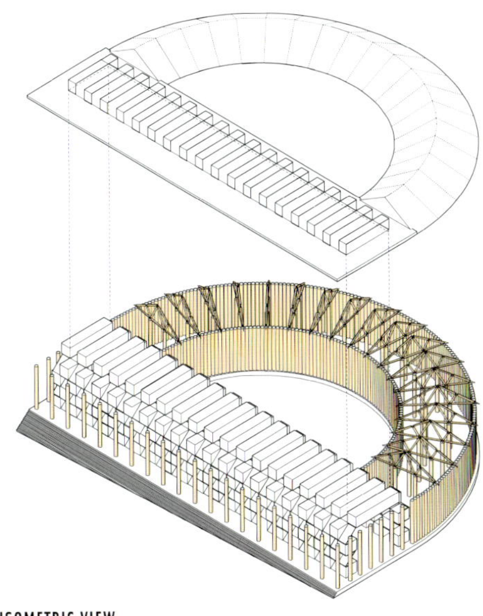

ISOMETRIC VIEW

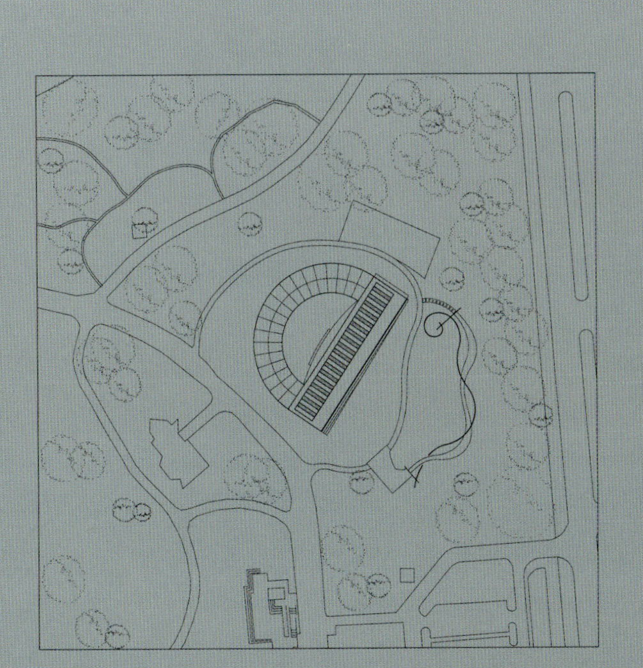

SITE PLAN

여자를 밝히다
브랜드를 밝히다
PAPER TAINER
PAPER TAINER

Fawood Children's Centre is a well thought out building designed especially for children's needs. Located in Harlesden, a run-down neighborhood in London, it is a ray of light for the entire area. It brings a nursery for three-to-five-year-olds, nursery facilities for autistic children and children with special needs, and a base for community education workers and consultation services. Soon, the Children's Centre that is currently still surrounded by tall concrete blocks will be sited within a new urban park in a revived neighborhood.

The Children's Centre has been designed to incorporate under one roof indoor as well as outdoor teaching and play areas. A colorful meshed shell and lightweight roof together cover and protect the open-air play surfaces as well as the classrooms and nurseries made from recycled shipping containers. As opposed to an average British nursery, typically housed in a one-level building surrounded by an open air play area that is, due to the climate, useless for most of the year, the Fawood children can use the protected external play areas all year round.

The project's shoestring budget called for the use of prefabricated and low-cost elements. Inside the shed-like enclosure, the necessary indoor areas (such as classrooms, nurseries, offices and service rooms) have therefore been placed into brightly colored containers, stacked into three-level clusters that resemble giant children's building blocks. The containers have their own lifts, stairs and underfloor heating, and are connected by decked walkways. One of the classrooms is in fact a Mongolian yurt – a much more magical place to be told a story than a conventional classroom, and yet another space inside the Centre that can help the children's imagination run free. At playtime children can enjoy as many as 14 different play facilities. They can chase each other through a "willow tunnel", have a picnic lunch in a "piazza", climb into a tree house, act on an outdoor stage, splash about in a water garden or swing from a climbing frame – independent of the weather and inside the protective steel-mesh walls. Such a safe environment, interspersed with colorful and imaginative play areas, the Fawood Children's Centre is surely a nursery that every child would love to call their own.

NAME ➔ **FAWOOD CHILDREN'S CENTRE**
USE ➔ NURSERY & ADULT EDUCATION CENTRE ARCHITECTS ➔ ALSOP DESIGN LTD LOCATION ➔ LONDON, ENGLAND, UK YEAR ➔ 2005
CLIENT ➔ STONEBRIDGE HOUSING ACTION TRUST

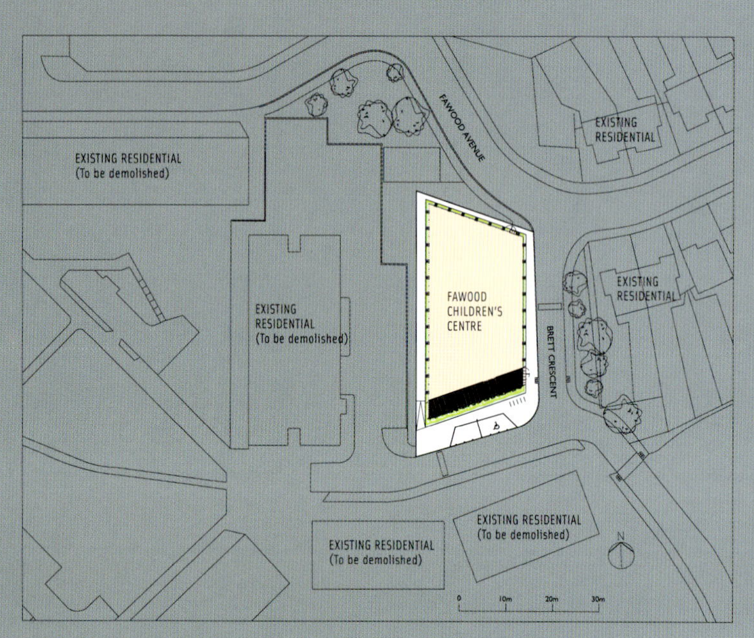

SITE PLAN

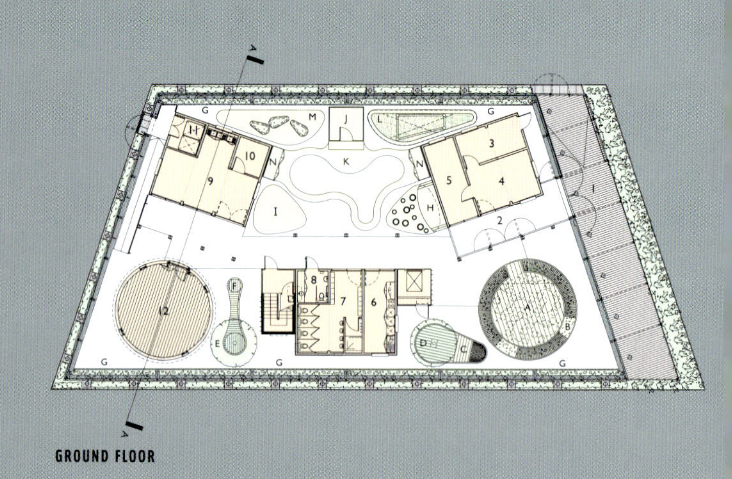

GROUND FLOOR

CROSS SECTION

F A W O

NAME TREEHOUSE TEMPORARY SCHOOL

USE SCHOOL ARCHITECTS PENOYRE & PRASAD LOCATION LONDON, ENGLAND, UK YEAR 2001 CLIENT TREEHOUSE TRUST

The TreeHouse Trust is an English charity, founded by a group of parents whose children had been diagnosed with autism. In 2001 they erected in London a temporary school for their children, manifesting their efforts to enhance the children's quality of life, while also being environmentally-conscious.

Every child's dream is to have a small shelter among the tree tops, no one's but their own, where they can escape into their own world. The tree-house temporary school symbolizes happy childhoods where this is possible, and shows care for children and for the development of their imagination. Both is reflected in the name of the TreeHouse Trust, as well as in the building that it erected and that goes up among tree tops – neither is incidental.

The two-level 600m² building consists of 36 prefabricated container units. It would be nothing but another container-constructed facility, were it not for its nature-friendly concept, even more striking due to the school's temporary character. It was erected on a site overgrown with protected London plane trees. There was no room for cranes, therefore a unique flat packed construction was designed to allow a quick and simple services fit out, preventing tree-root damage. The larch cladding thoroughly conceals the underlying metallic blue of the containers and blends in nicely with the surrounding trees. With its environment-friendly attitude, this temporary home for the unique parent-led Treehouse School has decisively set a new standard for this type of temporary construction.

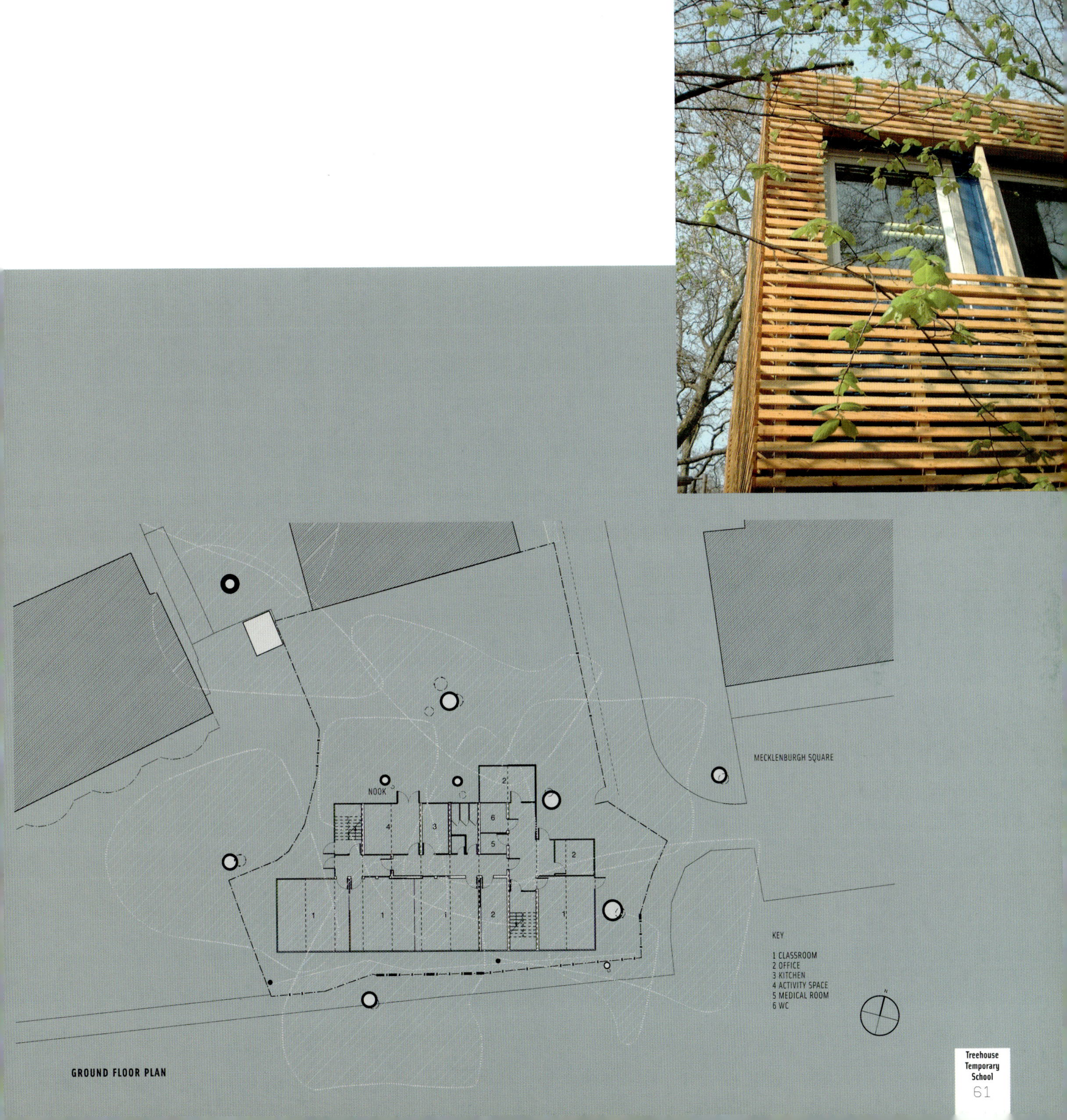
NOOK
MECKLENBURGH SQUARE
KEY
1 CLASSROOM
2 OFFICE
3 KITCHEN
4 ACTIVITY SPACE
5 MEDICAL ROOM
6 WC
GROUND FLOOR PLAN

NAME CHILDREN'S ACTIVITY CENTRE
USE ACTIVITY CENTRE ARCHITECTS PHOOEY ARCHITECTS
LOCATION MELBOURNE, AUSTRALIA YEAR 2007
CLIENT CITY OF PORT PHILLIP

Although some container structures try to beautify containers or even conceal them so as to make the buildings appear more conformist, Melbourne's Children´s Activity Centre by Phooey Architects goes back to the basics. Rusty patches and logos of shipping carriers on the facade remain clearly visible and are interrupted only by an occasional window opening. The four containers – along with the staircases, projecting roofs and large wooden terrace – have been arranged so as to give the impression of a stranded pirate ship in an amusement park, which the children no doubt appreciate. The little users of the pirate ship therefore feel perfectly at home there, also due to its composition, since they are very much inclined to recycle and make innovative use of available materials themselves (this is especially evident when they construct hiding places using every possible item close to hand).

Two larger containers sit on the ground and support the two smaller ones, rotated by 45° and thus overhanging the bottom structure. A large part of the bottom two containers' roofs is left bare to form the "ship's deck" – a spacious wooden terrace, which is connected to the ground by an external staircase. The staircase boasts an original decorum made of container steel leftovers from cuts for window and door openings. The fence and projecting roof on the first floor are of the same making. The balconies wedged between the two door wings of standard containers provide shade in summer, while winter comfort is achieved through orientation and ample insulation.

The interior is simple, with the largest space being the so-called flexible multiuse room on the ground floor. Lined with chequered carpet tiles, it is used for study, painting, dancing and lounging about. The "ship" is surrounded by trees, garden, pond and assortment of sandpits and play areas, which further creates a genuine adventure playground.

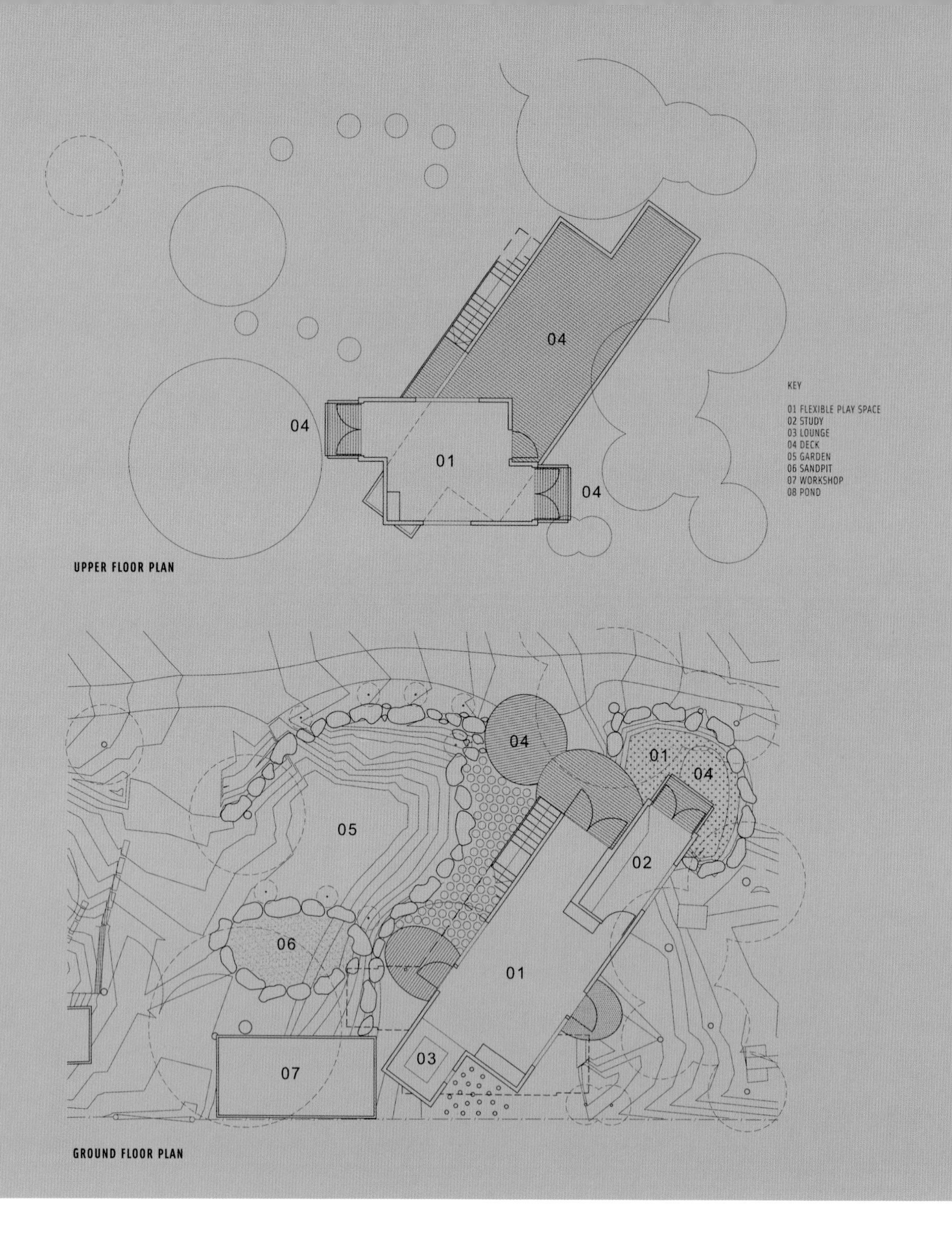
04
04
01
04
UPPER FLOOR PLAN
KEY
01 FLEXIBLE PLAY SPACE
02 STUDY
03 LOUNGE
04 DECK
05 GARDEN
06 SANDPIT
07 WORKSHOP
08 POND
04
01
04
05
02
06
01
03
07
GROUND FLOOR PLAN

MAEU 814 094 2
45G1
MAERSK

NAME➔**BED BY NIGHT**
USE➔SHELTER FOR STREET CHILDREN ARCHITECT➔HAN SLAWIK LOCATION➔HANNOVER, GERMANY YEAR➔2002 CLIENT➔HAUPTLANDMANSHAFT HANNOVER

As the name suggests, Bed by Night is a shelter for street children in the German city of Hannover. 14 refurnished and five new containers were remodeled to accommodate the necessary facilities for sleeping, guidance, administration, catering, residence and protection of street children. On the ground floor, there are the lobby, offices and conference rooms on one side, and the kitchen and dining room on the other; the latter make up the day zone, which faces the inner weatherproof courtyard. The living area created between the assembled containers comprises an animated space both to live in and to look at from the outside. The two-level air space inside the encasement compensates the narrowness of the container interiors.

As direct exchanges between the inside and outside world were to be prevented, the architect chose a translucent industrial façade, through which only a faint outline of the people and objects is discernible, ensuring complete privacy and protection to the young occupants. Sleeping quarters have been placed in the upper storey so as to give a heightened feeling of security, while the windows on the ground floor can thus stand open during the day as well as by night allowing the air to circulate.

The containers making up the protecting encasement are held in primary colors and thus easily localized from the nearby major road, drawing the attention of passers-by to the complex. This in itself is an important aspect, since street children are a part of our society and should not be hidden. On the other hand, a colorful composition does not cost any more than grey varnish paint. The containers inside the protecting encasement are held in secondary colors, giving the complex a gentler and softer appearance from close up. This is to humor the children inside, since architectural psychologists believe it is these colors that are most appealing to children.

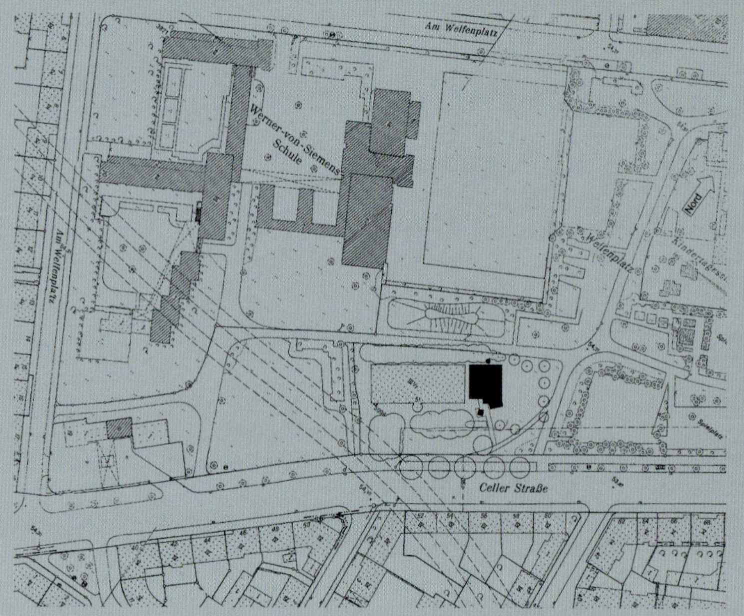
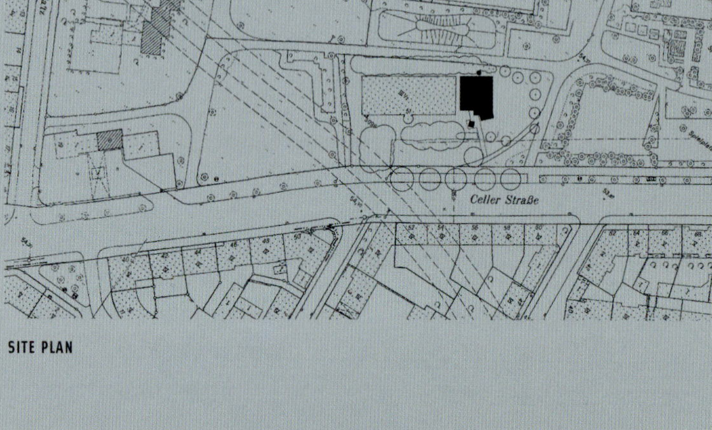

SITE PLAN

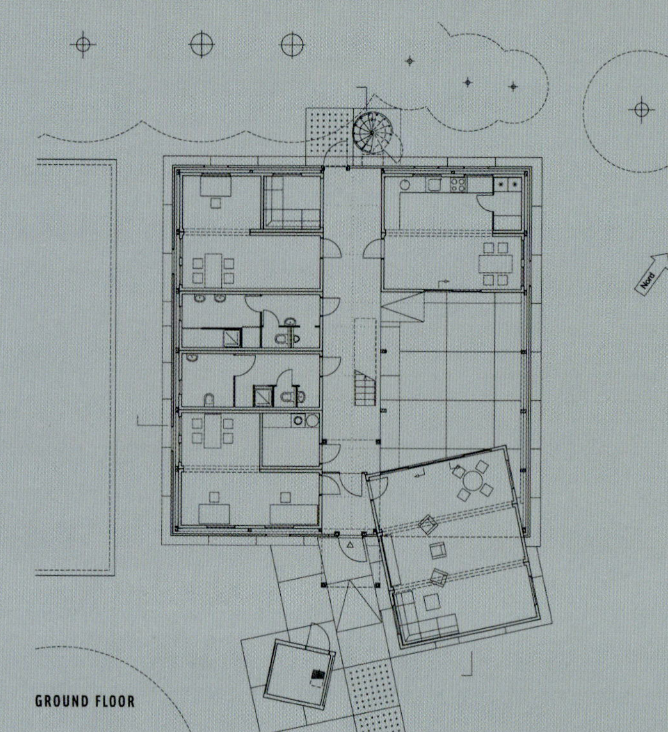

GROUND FLOOR

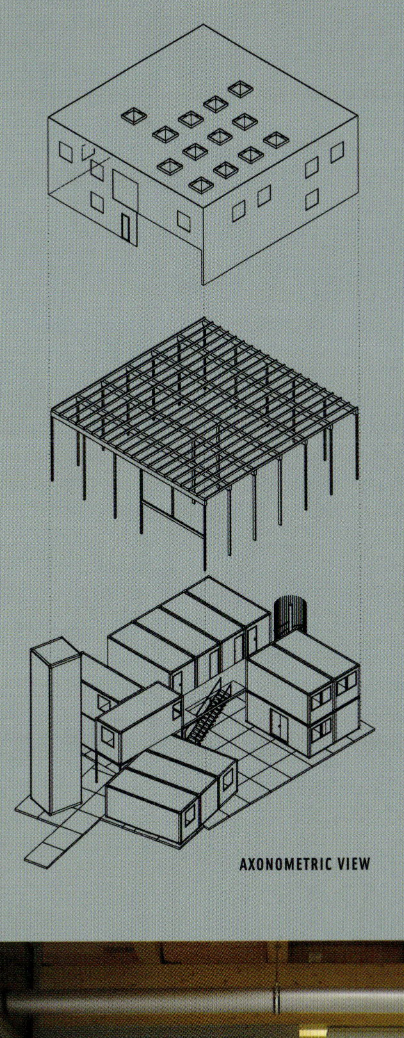

AXONOMETRIC VIEW

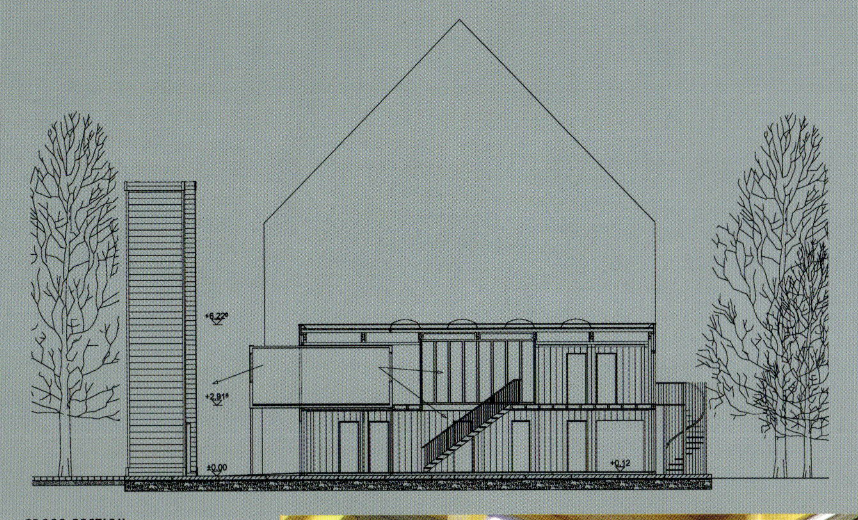

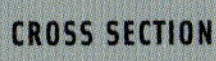

CROSS SECTION

NAME IJBZ SANITARY FACILITY

USE SANITARY FACILITY ARCHITECTS AFF ARCHITEKTEN
LOCATION BARLEBER SEE, MAGDENBURG, GERMANY
YEAR 2002 CLIENT YOUTH OFFICE MAGDENBURG

In Germany's Magdeburg area, where the River Elbe runs into the Mitteland Canal, several artificial lakes had formed in the 1930s and gradually developed into places of recreation and biotopes. The oldest of the lakes, the Barleber See lake, has long been a bathing beach and a camp site, while currently hosting the Jugend youth and community centre, with the guests being accommodated in bungalows and tents.

The architects were commissioned to design a replacement for the run-down sanitation facility in the camp, at the lake's northern brim. The approaching tourist season urged the replacement to come fast, and thus the client proposed a container sanitation unit. The AFF Architekten constructed it out of three eye-catching elements – standard containers, "crown roof" with roof windows, and entrance porch from an outer wooden lagging. The choice of color as well as added roof and entry-way details make the container construction stand out as a unique icon of its environment; small touches have given the sanitation unit an entirely new visual value.

Playful surfaces and colors make the construction attractive for children. The texture tells about what the house has experienced, whether it is friendly or unfriendly, what it is made of, and what you can do with it. To bring the building closer to its users, and to break the functional rigor of conventional sanitary containers, pink was chosen as the color for the outside shell, while associative wall patterns decorate the rooms inside. The patterns used in boys' and girls' changing rooms are similar, yet different in color and imagery, with blue sharks appealing to boys and pink dolphins to girls. The robust simplicity of the containers and their fittings contains a narrative component, which gives the children the freedom to pass on the story.

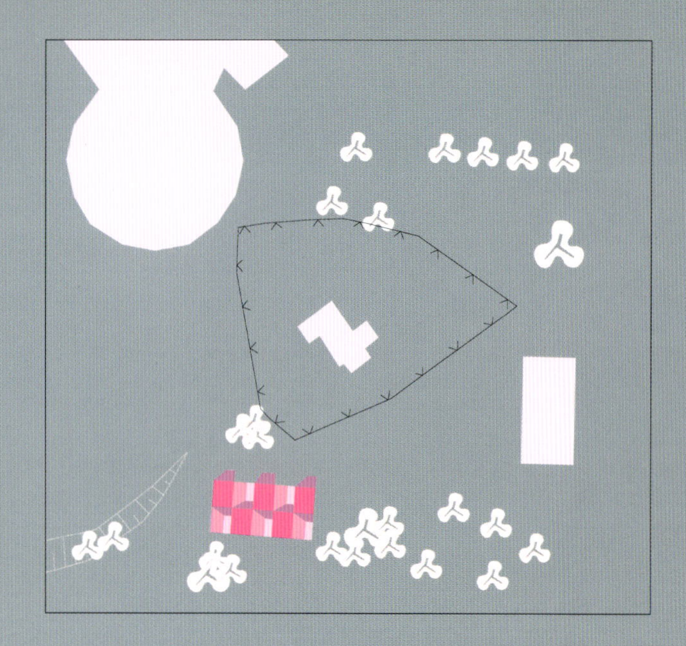

SITE PLAN

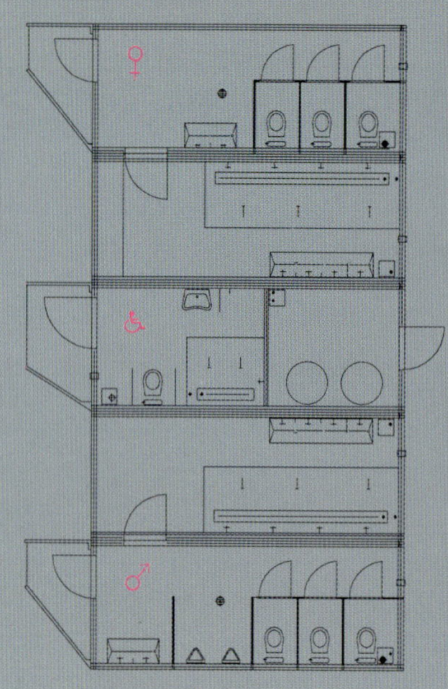

FLOOR PLAN

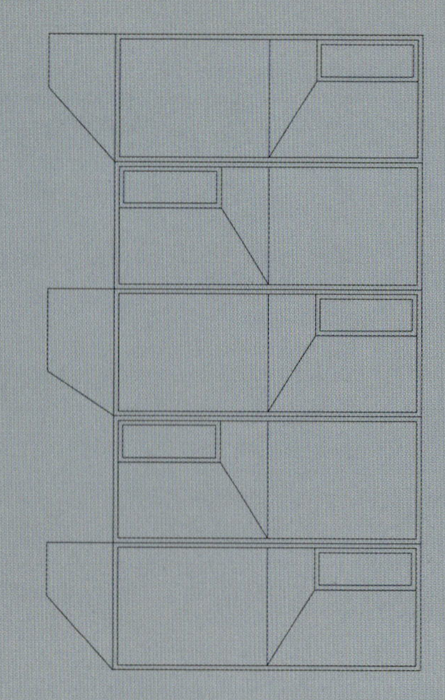

ROOF PLAN

3D MODEL

48 CONTAINERS

NAME HH CRUISE CENTER
USE PASSENGERS TERMINAL ARCHITECTS RENNER HAINKE WIRTH ARCHITEKTEN LOCATION HAMBURG PORT, GERMANY YEAR 2002 CLIENT HAFENCITY HAMBURG GMBH

RHW Architekten 78

Passengers docking in the Hamburg harbor after a cruise at sea set their foot ashore into the temporary HH Cruise Center terminal by RHW Architekten. In planning the passenger terminal, architects subsumed to two guiding themes – the two familiar elements of seafaring: the traditional overseas container, and the large white sail that symbolizes elegance, luxury and leisure.

The terminal's walls are composed of colored ship containers, bound on top by a sumptuous, magically illuminated roof facing the city. At night, the lit roof transforms into a hovering canvas and assumes the role of the fifth façade of the building. A wide glass opening in the façade, 'the window to the city', directs the view of the arriving cruise-ship passenger to the city skyline and onto the most famous church of Hamburg, the St. Michel's. The terminal's spacious central hall is where the city visually meets the harbor, and vice versa, and has great potential as a place for a variety of special events.

The atmosphere of the harbor will be further enhanced by a grid of vertically standing container follies around the cruise centre, which will shine at night due to a simply applied translucent foil and thus illuminate the parking area. These lighthouses on land will visually attract attention to the 'Hafencity' and give new orientation within the vastness of space.

Since the construction material was containers, the architects were able to erect, within the set budget, a building of larger dimensions and a more spectacular design than would be possible with traditional building techniques.

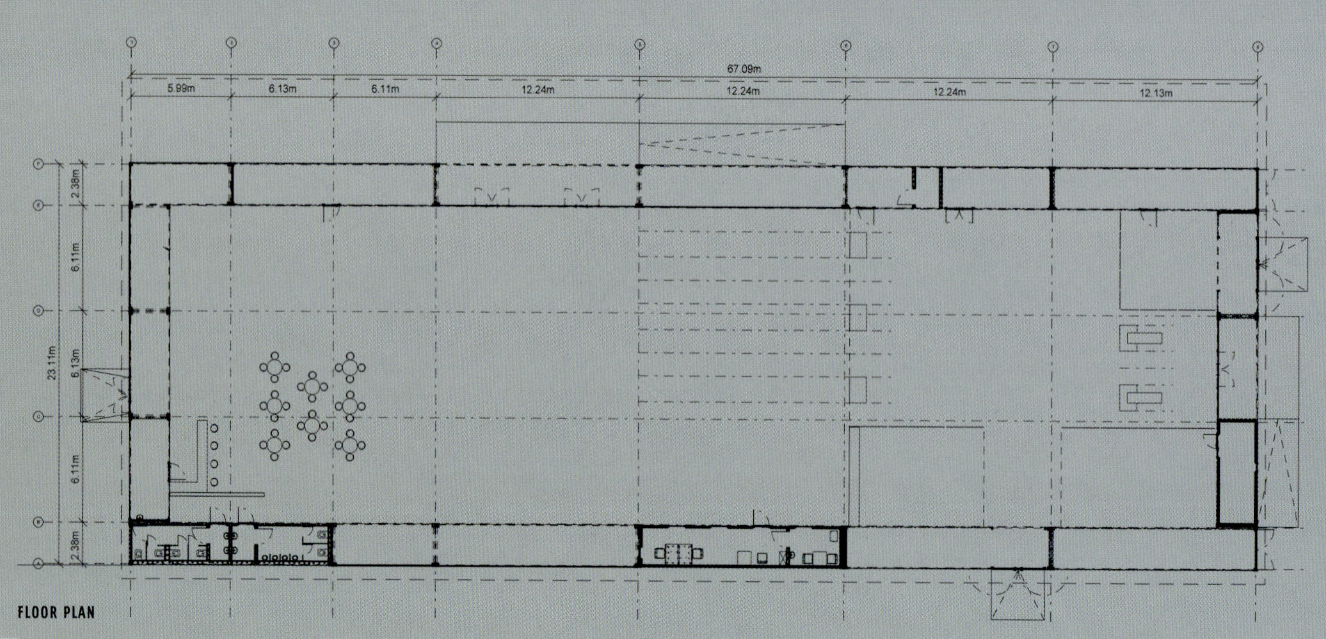
67.09m
5.99m
6.13m
6.11m
12.24m
12.24m
12.24m
12.13m
2.38m
6.11m
23.11m
6.13m
6.11m
2.38m

FLOOR PLAN

Hamburg
Cruise Center

Hamburg
Cruise Center

Hamburg
Cruise Center

24 CONTAINERS

NAME CENTRO TECNOLÓXICO RURAL
USE WORK/LIVE UNITS ARCHITECTS ARQUITECTOS ASOCIADOS DE SANTIAGO LOCATION MARROZOS, SPAIN YEAR 2006
CLIENT XUNTA DE GALICIA

In a small village on the outskirts of Santigo de Compostela, Spain, lies Centro Tecnolóxico Rural, a rural technological center. It resulted from research into new forms of living spaces, commissioned by the local authorities, who wanted to provi de people with decent housing at reasonable prices. These "espacios habitales" should preserve the environment, make use of alternative energy sources and be constructed from recycled materials, while also offer an opportunity to the local industry to participate.

The Center sits in the slope of a hill, offering fantastic green views of the hilly landscape below. It is composed of 24 containers, which were shipped in from the port in Vigo. After being cleaned, painted and insulated, they were stacked into a two-part composition. The lower part consists of six containers making up four individual units and a large glazed area between two of the containers, which is intended as a spacious office or classroom. The upper level consists of 18 containers making up eight individual units and the central area, the four-level tower. Each level of the tower is constructed from two containers standing apart and connected by a large glazed area, which is intended, like the one on the ground level, for spacious offices, classrooms or meeting rooms. The remaining containers house offices and workshops, or can be remodeled into living areas. Each unit consists of a 40' container with one shorter side glazed to allow light enter the room, and a balcony. The office complex also includes a garden, swimming pool, spa, gym and kitchen, offering all that it takes to work and live in a pleasant environment in the countryside.

The Center targets the young who would like to bring their post-university knowledge to the countryside to work, but need business premises to start with. The most wanted profiles include high-tech experts, but architects, economists, lawyers, etc are just as welcome. It is namely the young that can revive and enrich the countryside.

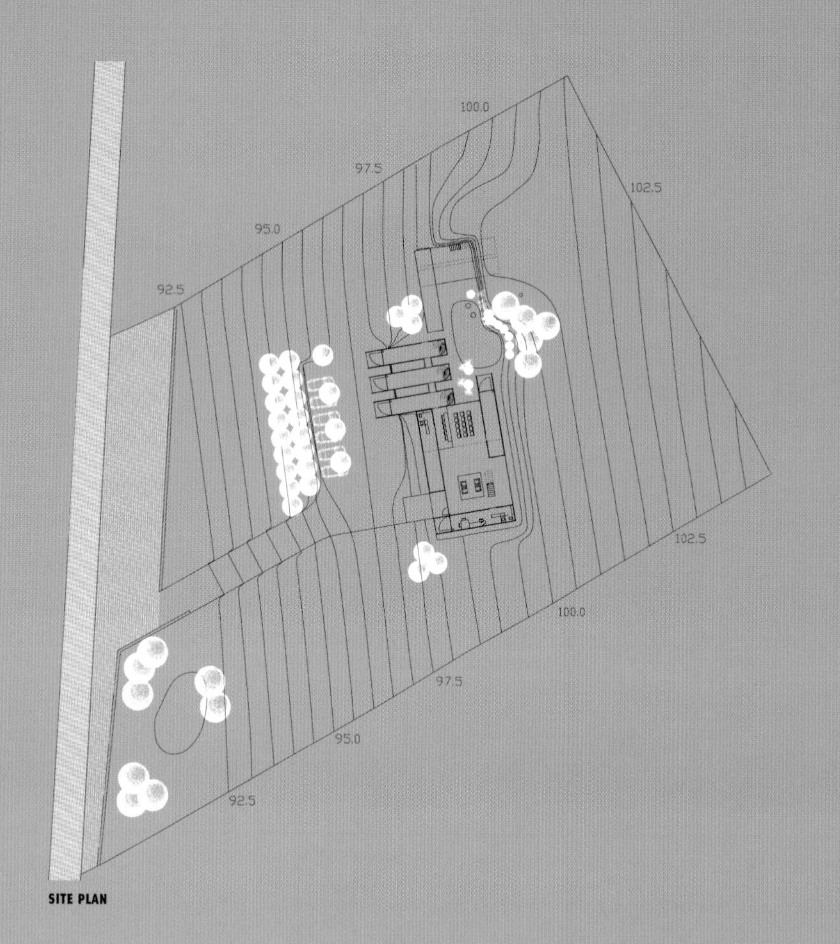

SITE PLAN

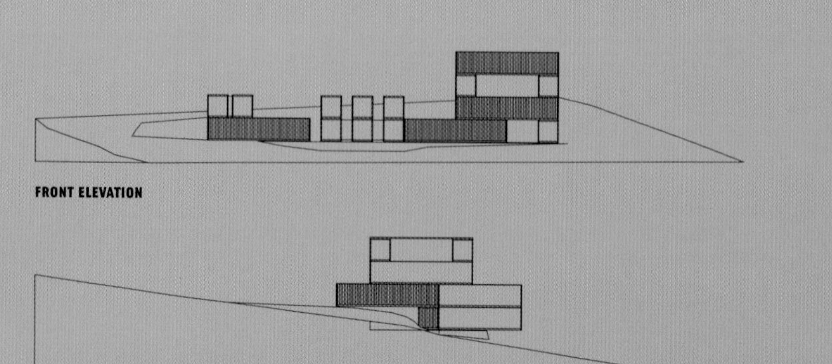

FRONT ELEVATION

SIDE ELEVATION

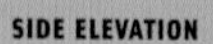

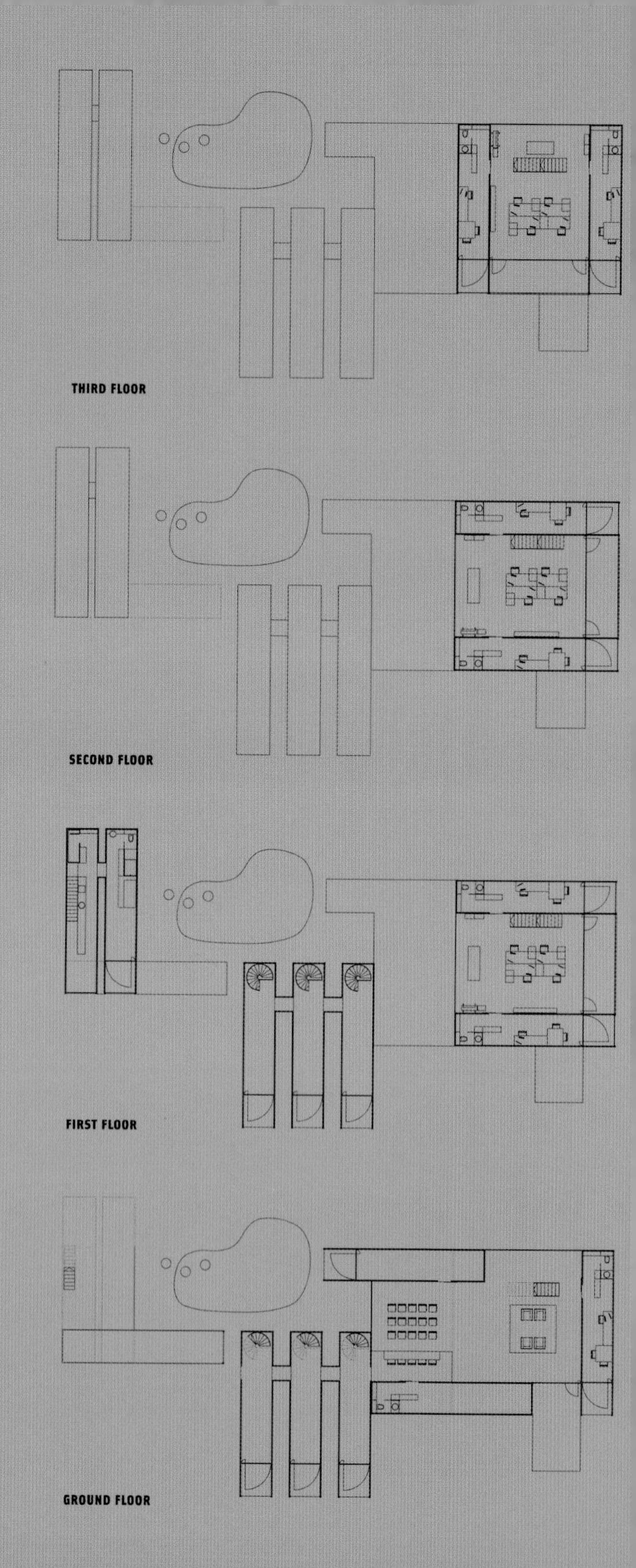

THIRD FLOOR

SECOND FLOOR

FIRST FLOOR

GROUND FLOOR

12
11
10

NAME FREITAG FLAGSHIP
USE STORE ARCHITECTS SPILLMANN ECHSLE
LOCATION ZURICH, SWITZERLAND YEAR 2005
CLIENT FREITAG LAB A.G.

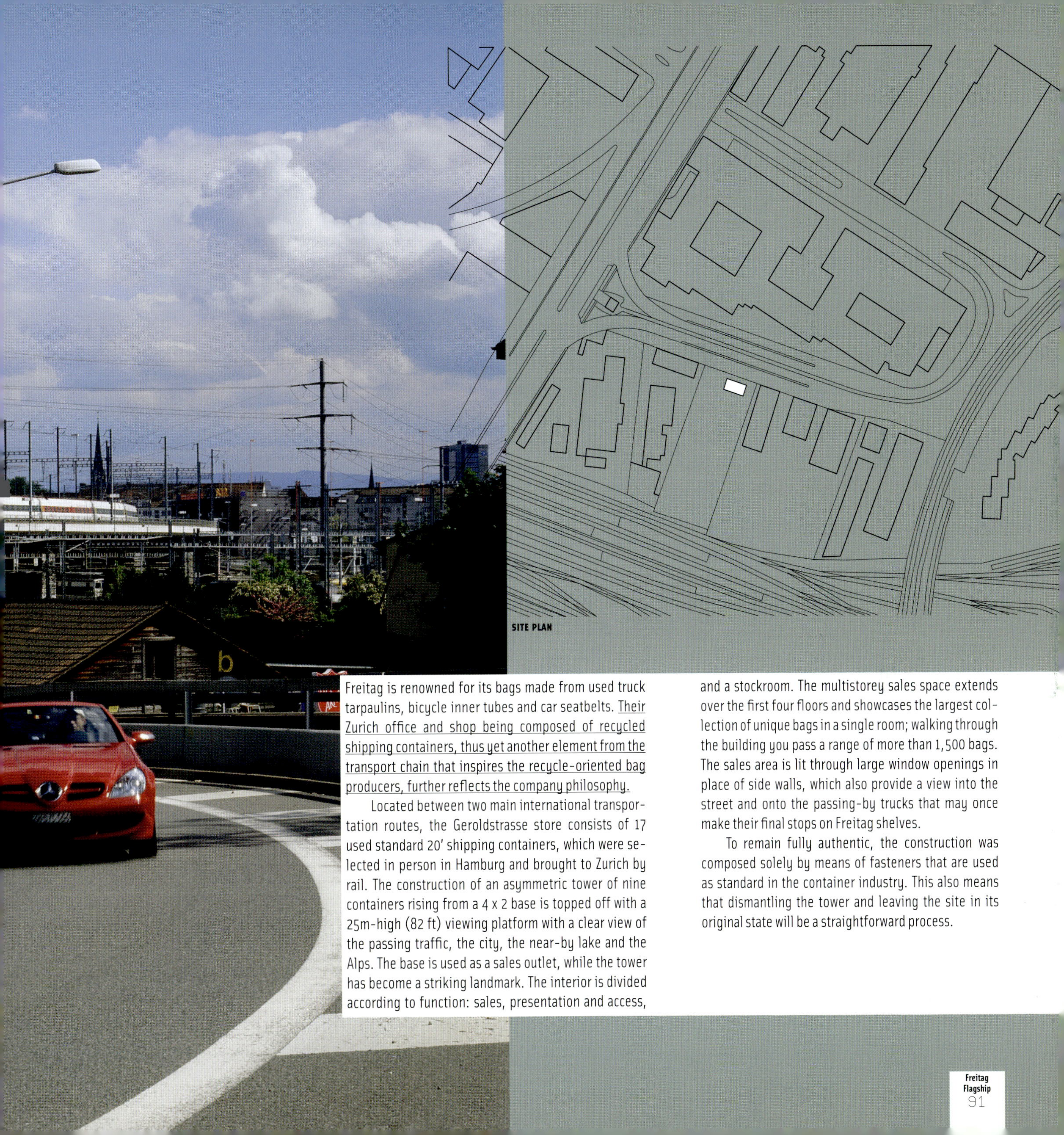

Freitag is renowned for its bags made from used truck tarpaulins, bicycle inner tubes and car seatbelts. Their Zurich office and shop being composed of recycled shipping containers, thus yet another element from the transport chain that inspires the recycle-oriented bag producers, further reflects the company philosophy.

Located between two main international transportation routes, the Geroldstrasse store consists of 17 used standard 20' shipping containers, which were selected in person in Hamburg and brought to Zurich by rail. The construction of an asymmetric tower of nine containers rising from a 4 x 2 base is topped off with a 25m-high (82 ft) viewing platform with a clear view of the passing traffic, the city, the near-by lake and the Alps. The base is used as a sales outlet, while the tower has become a striking landmark. The interior is divided according to function: sales, presentation and access, and a stockroom. The multistorey sales space extends over the first four floors and showcases the largest collection of unique bags in a single room; walking through the building you pass a range of more than 1,500 bags. The sales area is lit through large window openings in place of side walls, which also provide a view into the street and onto the passing-by trucks that may once make their final stops on Freitag shelves.

To remain fully authentic, the construction was composed solely by means of fasteners that are used as standard in the container industry. This also means that dismantling the tower and leaving the site in its original state will be a straightforward process.

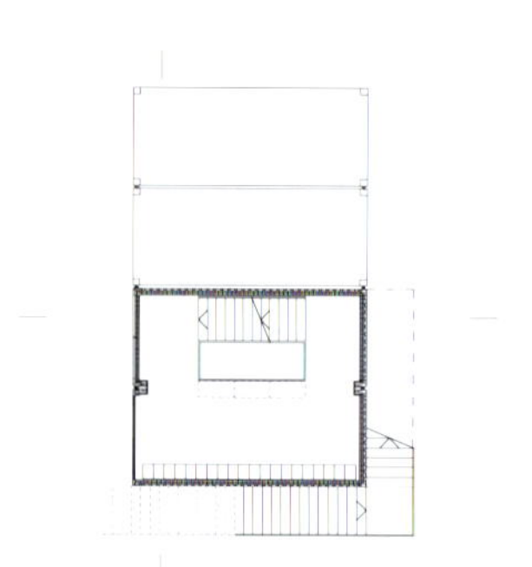

SECOND FLOOR PLAN

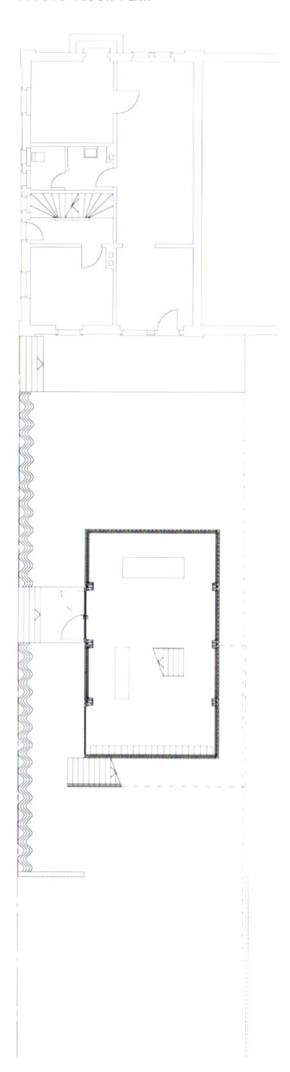

GROUND FLOOR PLAN

CROSS SECTION

LONGITUDINAL SECTION

FREITAG

FREITAG
SHOP
ZÜRICH
OOCL
TIPHOOK
CAI
CAI
TIPHOOK
OPEN NOW

SHOP
ZH
msc
OOCL
EVERGREEN
MAERSK

1
CONTAINER

NAME➔**UNIQLO**
USE➔MOBILE RETAIL UNIT ARCHITECTS➔LOT-EK
LOCATION➔NEW YORK, USA (MOBILE) YEAR➔2006
CLIENT➔UNIQLO USA INC.

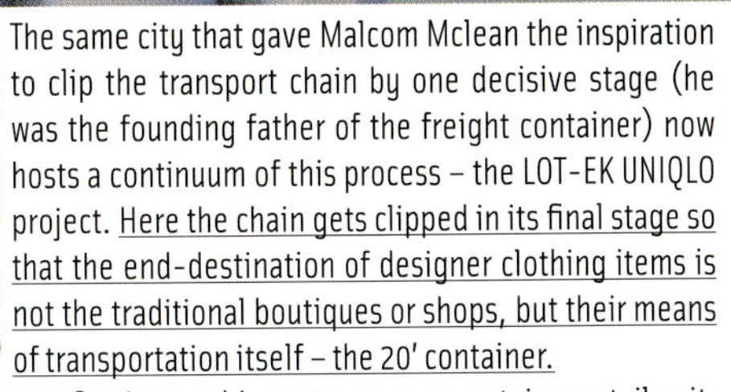

The same city that gave Malcom Mclean the inspiration to clip the transport chain by one decisive stage (he was the founding father of the freight container) now hosts a continuum of this process – the LOT-EK UNIQLO project. Here the chain gets clipped in its final stage so that the end-destination of designer clothing items is not the traditional boutiques or shops, but their means of transportation itself – the 20′ container.

Setting up this contemporary container retail unit, the goal of the UNIQLO company (UNIQue cLOthing) was to attract attention and inform buyers of their entering the American market. A striking advantage of such a store is its mobility, since it can be placed into any part of the city and can temporarily even squat locations where its long-term erection would not prove economically viable. Surely the very outlook of a cargo container stirs consumer interest everywhere it appears and thus automatically markets its merchandise.

The shop's design is simple. The single-color exterior carries the company's logo, with the minimalistic interior being furnished with plywood. One of the longer sides is covered in shelves of clothes top to bottom, while the parallel side contains a disabled-friendly entrance and a series of narrow windows filling the room with light. The shorter sides are both mirrors that optically expand the relatively small interior. The container shop also has a futuristic "changing tube" made of spiral wire and cloth, which comes down from the ceiling when people want to try on clothes and then goes back up to save space.

EXIT
UNI
QLO

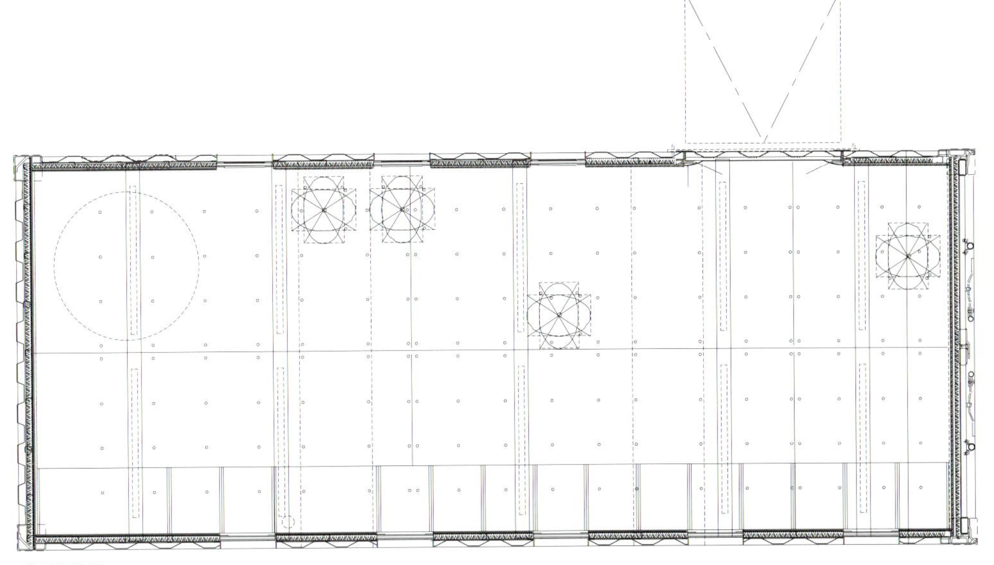

FLOOR PLAN

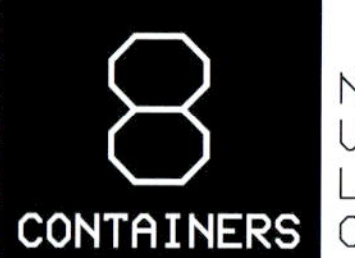

NAME➔**WIJN OF WATER RESTAURANT**
USE➔TEMPORARY RESTAURANT ARCHITECTS➔BIJVOET ARCHITECTUUR
LOCATION➔ROTTERDAM, THE NETHERLANDS YEAR➔2005
CLIENT➔BEA OOMENS

It took but six months from the first sketches for the Wijn of Water restaurant on the banks of Rotterdam's Maas River to open its doors. It was set up there for the period of two years, until it can move to its final venue close by. Given the limited budget and temporary nature of the project, shipping containers lent themselves as the only logical solution for this short-lived accommodation.

Eight 40' light blue containers were assembled into a three-side atrium, the back side of which rises in two levels. The fourth corner of the atrium is the standing ninth container, which towers 12m (39 ft) above the restaurant and serves as a landmark symbol from afar, while also being used as a handy storage. Containers were arranged so as to shelter guests on the terrace from northerly winds, while a canvas roof stretched over them shades the outside sitting area in the summer time. The ground floor of the restaurant houses the kitchen, lounge area, toilets and the restaurant itself, while on the first floor there are the staff room and installation room. All installations were taken outside the containers and stored in a metal tube on the roof, so as to make full use of the inner height of the relatively low containers and thus make the interior seem more spacious.

The restaurant sits 50 people. Guests enjoying the pleasant atmosphere can indulge in culinary delights by gazing out the large windows onto the River Maas. The expressionless two-level container wall facing the city acts as an antipode to the views that offer themselves to surprised guests once they enter the restaurant and catch a glimpse of the passing ships.

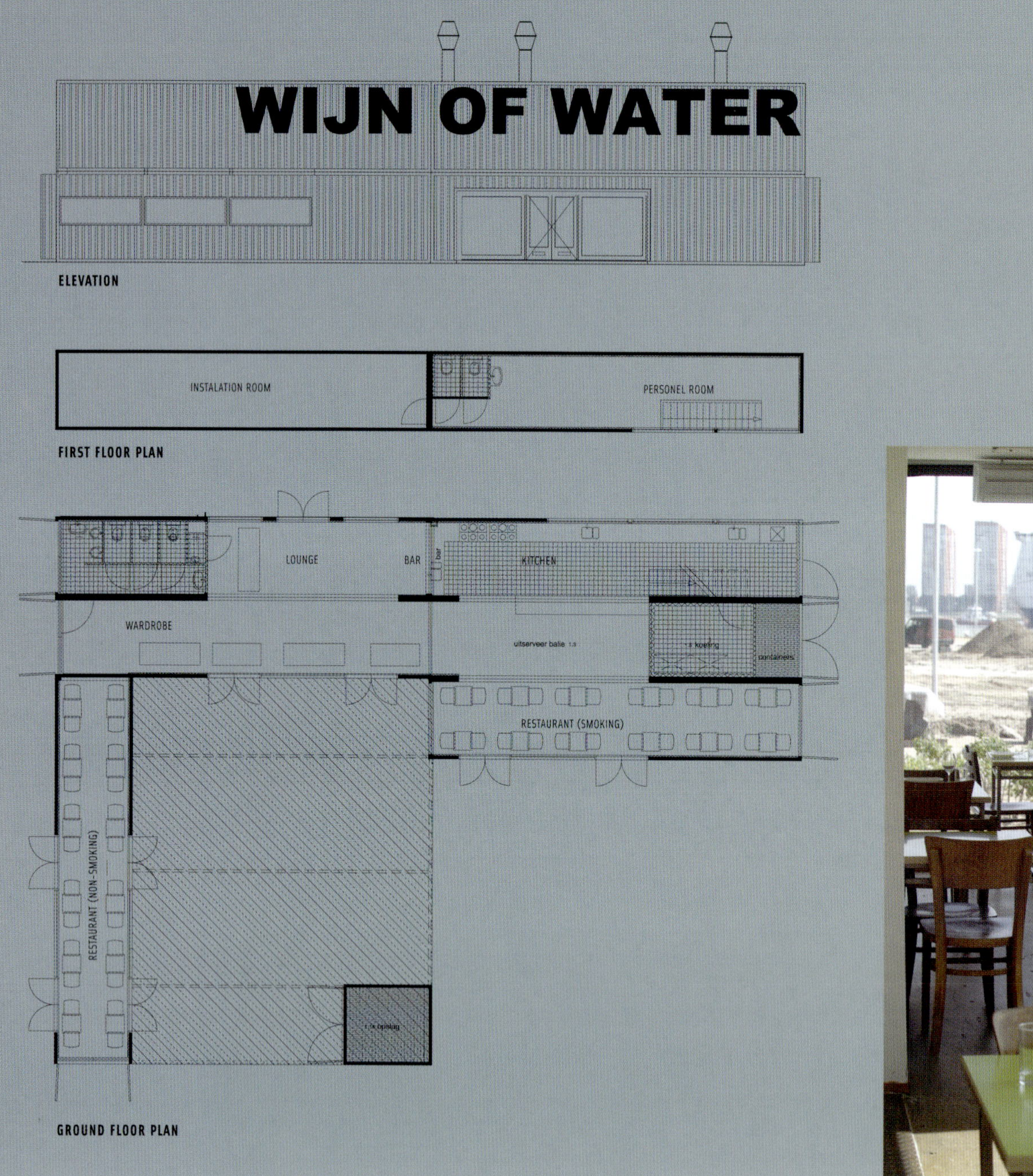
WIJN OF WATER
ELEVATION
INSTALATION ROOM
PERSONEL ROOM
FIRST FLOOR PLAN
LOUNGE
BAR
KITCHEN
WARDROBE
uitserveer balie
RESTAURANT (SMOKING)
RESTAURANT (NON-SMOKING)
GROUND FLOOR PLAN

P&O

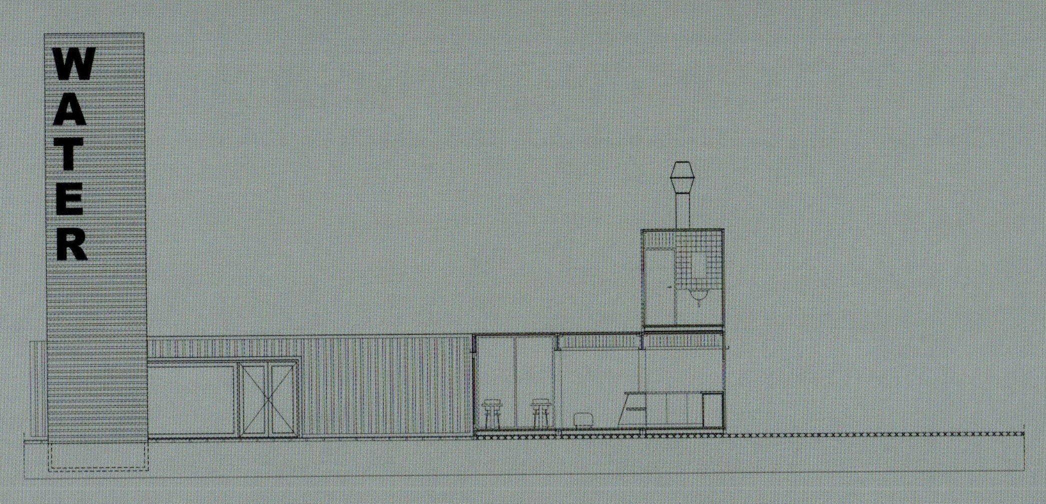

LONGITUDINAL SECTION

P&O

54 CONTAINERS

NAME ➔ **VOLVO C30 EXPERIENCE PAVILION**
USE ➔ PRODUCT PRESENTATION LOUNGE ARCHITECTS ➔ KNOCK.SE
LOCATION ➔ GOTHENBURG, SWEDEN YEAR ➔ 2006 CLIENT ➔ VOLVO CARS

Knock.se

Every two years Volvo invites its car dealers from all over the world to discuss the company's view of the future and, above all, to rally their enthusiasm for the cause. The 2006 event, which included the set-up of a temporary presentation spot for the company's new product, the C30, was entrusted to the Swedish 3-dimensional communication agency KNOCK. The event was staged in an old boat shed in a rural harbor outside Gothenburg, where they created an event hall from run-down shipping containers. The guests were transported to the venue by boat, only to arrive at an area and in front of a building that was nothing like they had expected to see.

Inside the shed, Knock created an environment where roughness and modern design where in contrast to each other, juxtaposing designs, colors, materials and tactile. The base shell of the container structure consisted of 54 somewhat rusty 20' transport containers. Nine rows of colorfully jumbled containers stacked three levels high on each side made up the longer sides of the event hall, while the two shorter sides served as 9 x 5m projection screens airing a multimedia presentation of the C30. The faded run-down shipping containers contrasted against 154 white-leather lounge sofas placed on a glossy white floor in the center of this temporary construction. A 150m-long white conveyor belt meandering through the room served culinary delicacies, which alternated with C30 target group lifestyle products.

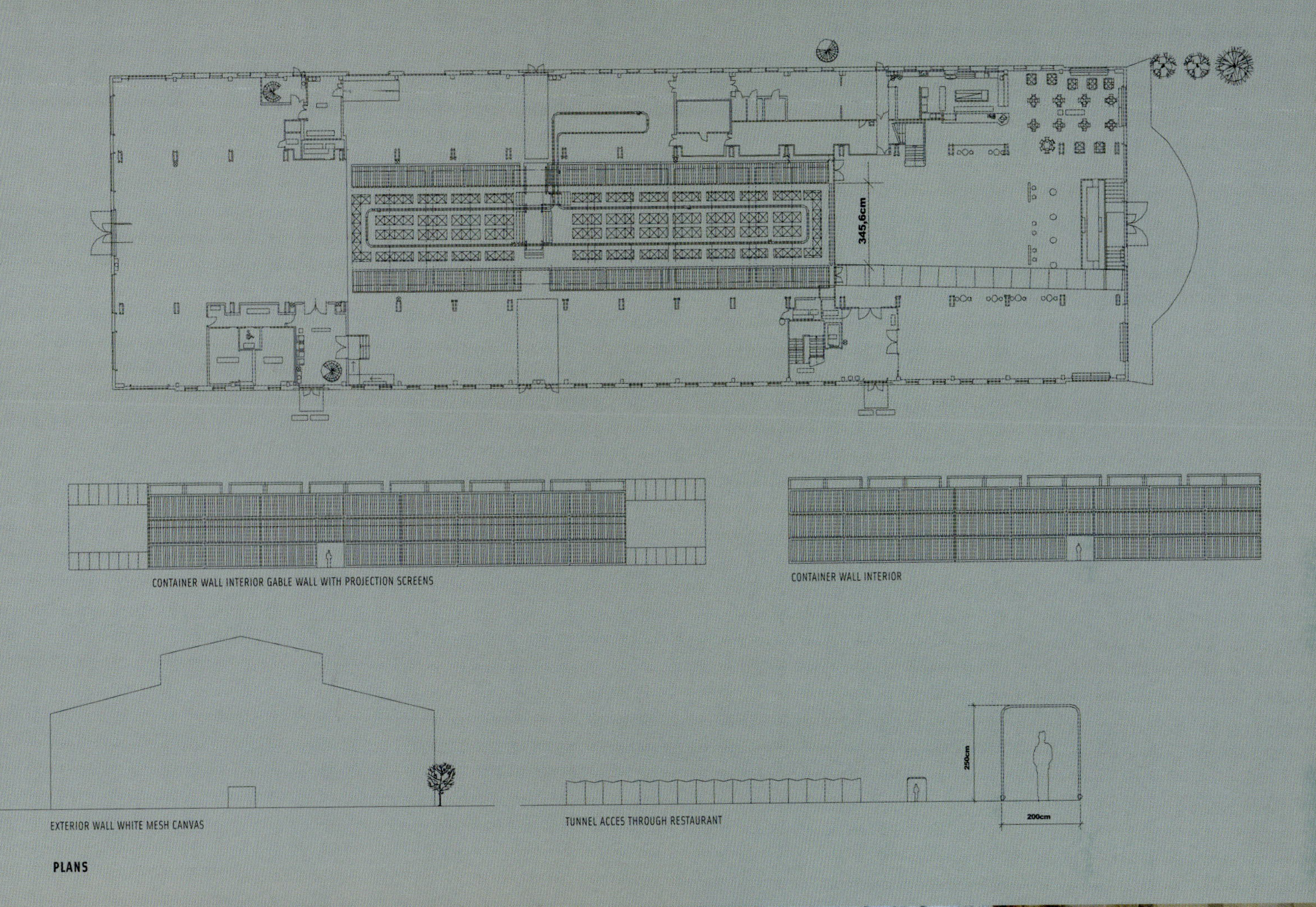
345,6cm
CONTAINER WALL INTERIOR GABLE WALL WITH PROJECTION SCREENS
CONTAINER WALL INTERIOR
250cm
200cm
EXTERIOR WALL WHITE MESH CANVAS
TUNNEL ACCES THROUGH RESTAURANT

PLANS

ic
61

NAME EXPO ACUEDUCTO
USE EVENT STAGE CONCEPT DESIGN: CORPORACTIVA ARCHITECTS BOPBA ARQUITECTURA LOCATION BARCELONA, SPAIN AND OTHERS YEAR 2007
CLIENT CORPORACTIVA S.L., EDELMAN SPAIN S.A. & EXPO ZARAGOZA

The planned international exposition EXPO 2008 will be hosted by Spain's fifth largest city, Zaragoza. Placed in a meander of the river Ebro, the exposition will evolve around the main theme of Water and sustainable development. Several emblematic buildings will be erected at the Zaragoza exposition site, evoking water-related images (such as the Water Tower, representing a drop of water), but not all EXPO 2008 venues will be static; BOPBA architects and Corporactiva have created for this purpose a special container compound of a mobile character. Looking like an aqueduct – the advanced and technically accomplished water supply facility dating back from the Roman era – it is designed to travel around different Spanish cities, spreading word of the coming exposition. Through its form and building blocks it is made of, it symbolically calls attention to both the increasing challenge of water supply (aqueduct) and the need for sustainable development (recycled containers).

The monumental "aqueduct" of a size comparable to its Roman predecessor consists of 13 (10', 20' and 40') containers. It can be erected inside larger emblematic buildings or outside, in prominent public spaces across cities, to help emphasize the importance of the event. It consists of three main parts; the portal, exposition place and event space. The portal gives the structure its predominant aqueduct shape and this is where many events take place. It includes three columns of three containers each, stacked into three levels, covered by two larger containers on top. One of the stacked containers houses the information office, while the others have screens and micro stages for musicians and other artists. The top two containers serve as a projection screen and this is also where trapeze artists descend from spectacularly during the show. The exposition place on the ground takes the shape of two letters L, where projections screens, posters, interactive photography and models all present the upcoming EXPO 08 and invite visitors to Zaragoza. The aqueduct further includes an open-air event space, which regulates the number of visitors by being partly enclosed by containers. Offering chill out sofas where people can relax, it is intended to host gatherings, projection shows and gastronomic events.

The mobile structure from recycled, environment-friendly containers can easily pack up and move to another location, causing virtually no site impact. With its striking appearance, powerful agenda and conceptual links to the EXPO, this multi-purpose mobile container aqueduct makes a true ambassador to this grand event.

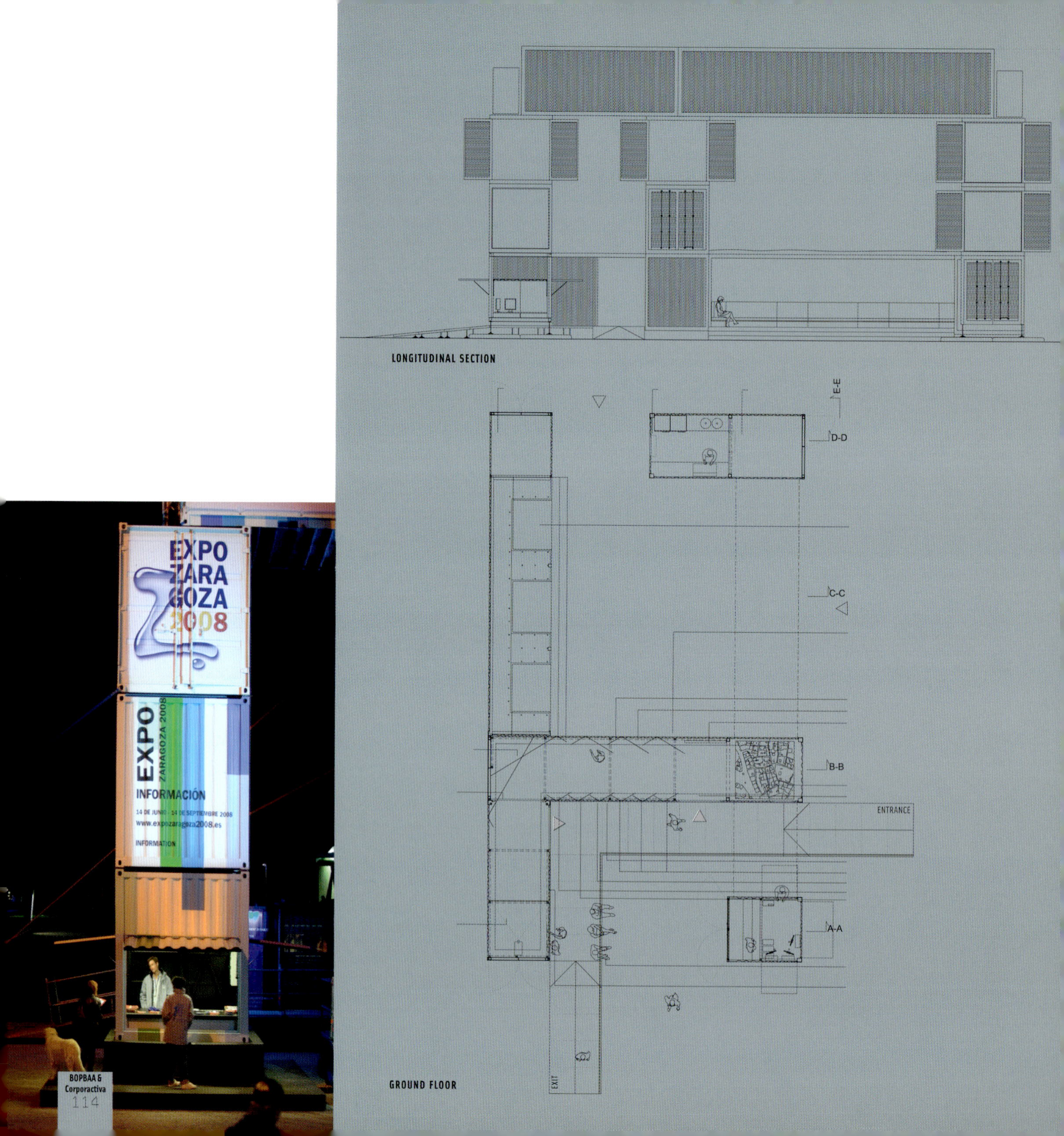
LONGITUDINAL SECTION
E-E
D-D
C-C
B-B
ENTRANCE
A-A
EXIT
GROUND FLOOR
EXPO
ZARA
GOZA
2008
EXPO
ZARAGOZA 2008
INFORMACIÓN
www.expozaragoza2008.es
INFORMATION

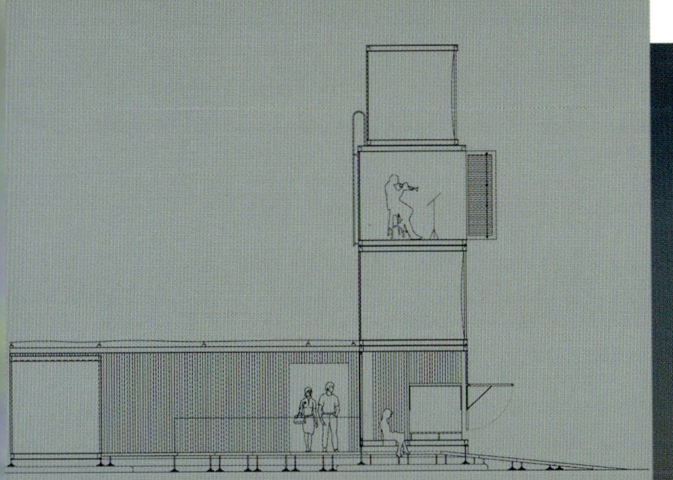

ROSS SECTION

EL MAYOR ESPECTÁCULO

Bélgica Belice
Bulgaria Cabo Verde
Chile China Chipre
Colombia Corea Costa Rica Croacia
Dinamarca Dominica Ecuador
EXPO ZARAGOZA 2008

CHAPTER
FIVE
PRESENTS
FOURTEEN
HOUSING
PROJECTS.

The range of container housing is wide and spans from small-size retreat cabins to container villas and large apartment buildings. Its history has seen three main stages. First there were attempts how to fit an entire apartment into a single steel box. Such apartments were designed mainly for the so-called urban nomads – the "side-products" of the modern society, however inhumane this may sound. Then there came smaller holiday houses created and owned mostly by architects and designers who thus paid tribute to this simply clever idea of living in a container. These are the people with a refined taste for what is trendy and containers are currently very "in". Container housing has become more widely acceptable for the public in the third stage, when containers are being combined with other construction materials, the result of which is houses that are quite similar to other custom housing architecture and containers only spice them up a bit.

Small-size container houses are still largely client-oriented and custom-made. The construction of container-made apartment buildings, however, is driven mainly by their practical value and economic efficiency, both for investors and users. The largest of container-made apartment buildings comprise as many as 1,000 units. The modular monotony of such a vast number of identical elements is broken down by diverse facades and installation patterns.

Of all the housing containers in urban environments, the one making up the Guzman Penthouse boasts by far the most prestigious location. It rests on a roof right below the Empire State Building on Manhattan, NY. Being light and prefabricated, containers are extremely well-suited for such elevated locations, where they can be deposited by a helicopter in a single day.

The Guzman Penthouse was created through the transformation of a mechanical room on the roof of a block. The south wall of the existing structure was replaced by a now all glass back side of a container, creating a bay window and letting more light into the room. The main level hosts an open layout of living/dining/kitchen, which thus receives plenty of light, along with a separate bedroom for the child and a bathroom. The longitudinal wall of the former machine room constitutes the visual and functional spine of the penthouse. Five openings pierce this wall and stretch the interior space towards the outside. Each opening accommodates a module of a different function, such as reading, watching, listening, while one also has the refrigerator.

This remodeled existing structure supports a 20' container, a half of which was "peeled off", so that the area that remained closed now hosts the bedroom, while the open part was made into a wooden patio. An internal steel fire escape ladder leads from the living area up into the bedroom, where a bed on tracks moves in and out of the closet, to make it a sleeping or lounge area, as needed.

To maintain the spirit of the original mechanical room, steel pipes and beams on the main floor have been left exposed.

1 CONTAINER

NAME→**GUZMAN PENTHOUSE**

USE→PENTHOUSE ARCHITECTS→LOT-EK LOCATION→NEW YORK, NY, USA YEAR→1996 CLIENT→PRIVATE

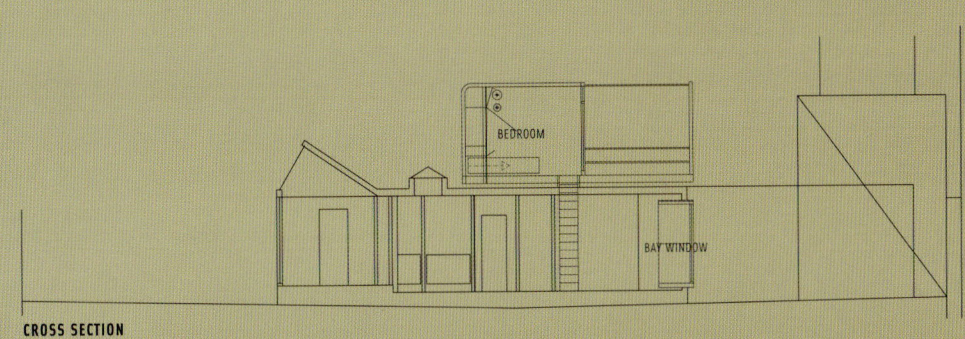

CROSS SECTION

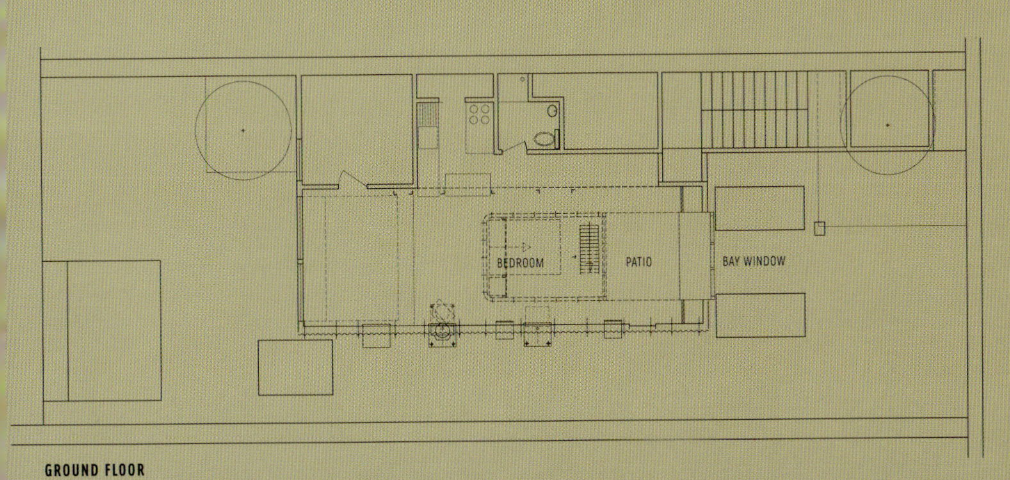

GROUND FLOOR

NAME ➔ **CUBES**
USE ➔ LIVE/WORK SPACE ARCHITECTS ➔ USM LTD
LOCATION ➔ PEATON HILL, SCOTLAND
YEAR ➔ 2002 CLIENT ➔ COVE PARK

USM Ltd

Cove Park is a centre for established artists, situated on the west coast of Scotland on 50 acres of spectacular countryside. It supports the work of artists in residence, offering them a place of retreat and providing them with time, space and freedom to undertake significant research, develop their practice and ideas and/or produce new work. The centre encompasses several accommodation facilities, including six en-suite accommodation units, known as cubes, composed from the 20'container.

As opposed to other projects which tend to spotlight the ISO container itself, this piece of container architecture stems to blend in with its rural surroundings. The cubes have been installed on a site overlooking Loch Long, an area of outstanding natural beauty. The grass-covered roofs make the cubes' rear parts appear as if grown out of the slopes, with the glass fronts catching a reflection of the pond below and vice versa. The cubes have a dark green exterior with round windows typical of Container City™ architecture.

A cube unit is composed of two containers and has a kitchen/sitting area and a shower room. They all offer double-bedded accommodation, except for one that has been fully equipped for the disabled and is the only cube with a single bed. Due to the all-glass front sides, the inside gets plenty of light, which is reinforced by the white interior that contrasts vividly with the cubes' colorful natural surroundings. Sliding glass doors lead out onto the decked balcony that runs along the front facade and extends over the pond with beautiful views of Loch. The cube units are connected/separated by means of standard container doors, enabling their residents privacy and independence.

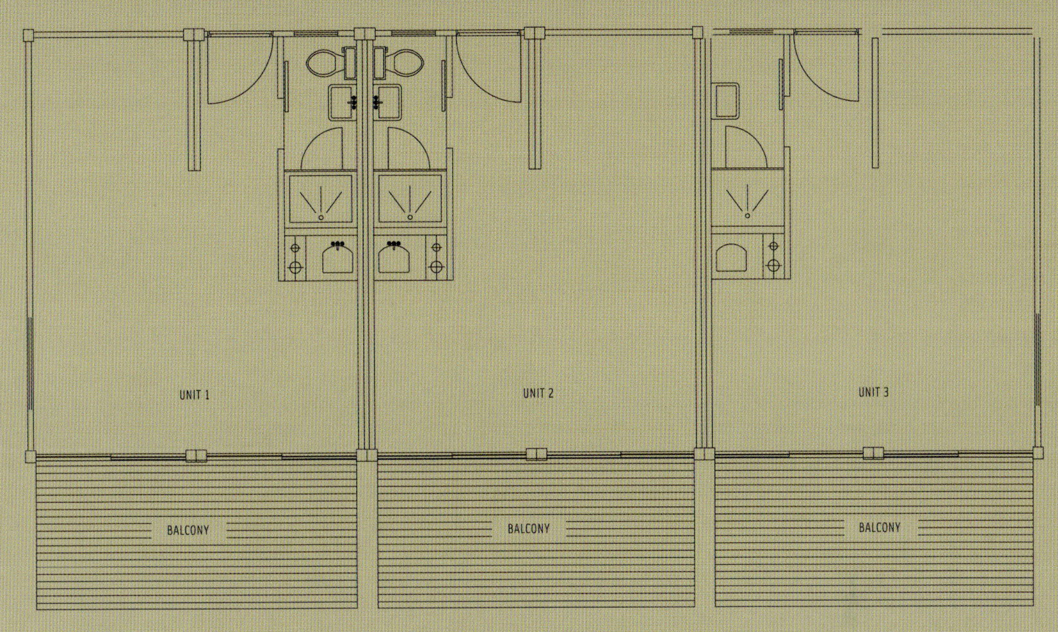

FLOOR PLAN

NAME C320 STUDIO

USE WEEKEND STUDIO ARCHITECTS HYBRID SEATTLE
LOCATION ENUMCLAW, WA, USA YEAR 2004
CLIENT RONNIE ALEXANDER

Sited on a riverside plot outside Enumclaw, Washington, C320 – an efficient cargotecture home – was set up as a rural vacation cabin. The two-container unit can house one person or a couple comfortably, and has been designed to be entirely off the grid.

The two containers are joined side-by-side and offset by 1.8m (6 ft), so that the interior space includes a large central area and two separate niches 1.8m-wide. The offset parts hold the bedroom and bathroom, while the living room and dining area are in the center. The inside is very bright – a shorter wall has a floor-to-ceiling window, while one of the longer sides is in fact a large sliding glass door, which extends the interior onto a wooden patio outside, and is closed up with outer-shell sliding container doors. The studio's fifth facade is its green fern-based roof, which makes it blend in with the surrounding nature if looked at from a bird's perspective.

The inside walls are covered in plywood, partly even in used plywood, contrasting with the cold steel exterior. Except for the window openings, the two containers have been left untouched, with original industrial orange paint and tracking numbers on the facade, testifying to their previous lives. As the architects say, the point of the cargo studio is not aesthetics, but the fact that the ideas driving it can change society. Although an oddity at first sight, contrasting with the trees and nature's softness, this is in fact an environmentally-friendly and fully autonomous unit. It uses propane and solar panels, and is fully covered in spray-in foam insulation. All used materials have been chosen so as to minimize toxic off-gassing. Site impact is virtually nonexistent, since Studio 320 could be unbolted in an hour, with the site being cleared in a day and the land returning to its natural condition in a matter of a single season.

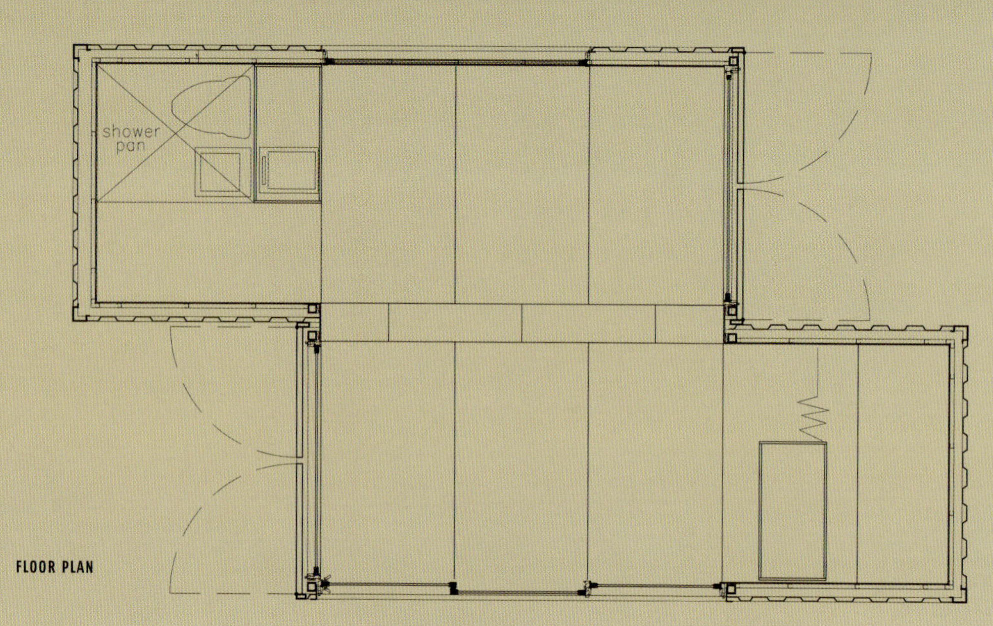

FLOOR PLAN

3960 42 7
25G1
MAX.GROSS
TARE

42 7
30.480 KGS.
67.200 LBS.
2.400 KGS.
5.290 LBS.
28.080 KGS.
61.910 LBS.
37.4 CU.M.
1.322 CU.FT.
CARGOTECTURE.COM

NAME STANKEY CABIN
USE WEEKEND HOUSE ARCHITECTS PAUL STANKEY LOCATION HOLYOKE, MI, USA
YEAR 2006 CLIENT STANKEY FAMILY

Located in the prairie some two hours north of Twin Cities, Minnesota, in a very small community called Holyoke, the Stankey container cabin is an exemplary DIY-house project employing shipping containers. A trailer home stood on the same spot years ago, but time took its toll, forcing its owners to search for a replacement. They looked for something nature-friendly and safe that could handle the wild seasonal changes and the intolerable rodent problem. After considerable research, the solution came in the form of cargo containers.

The cabin consists of two 20' used containers set apart and connected by a two-level glazed space with a lean-to roof. Being jacked onto piers, the house is elevated from the ground and has a staircase leading up to the entry into the central glazed area. The same size as the old trailer, the container cabin hosts a kitchen, dining room, living room, wash and clothes area, and two queen beds.

While the containers have no openings, the central area consists exclusively of glass surfaces, giving the interior plenty of light and spectacular views of the landscape. The interior walls and ceilings were finished in birch plywood. In this way the new Stankey cabin maintains the previous trailer's original feel, although in a new disguise. The cabin home will be powered, at least partially, by a small solar array, with the drainpipe being hooked up to a cistern to collect rainwater.

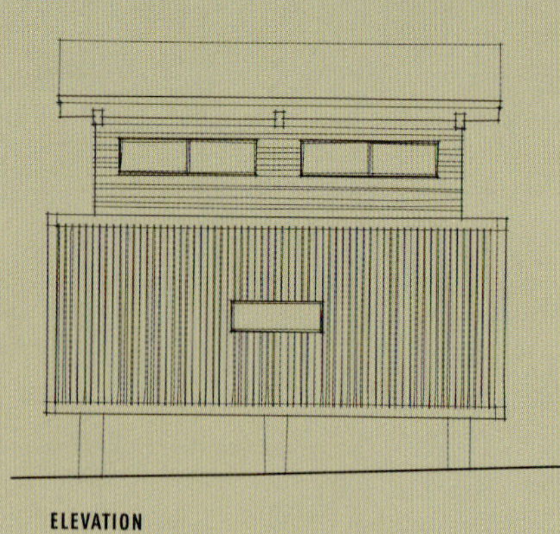

ELEVATION

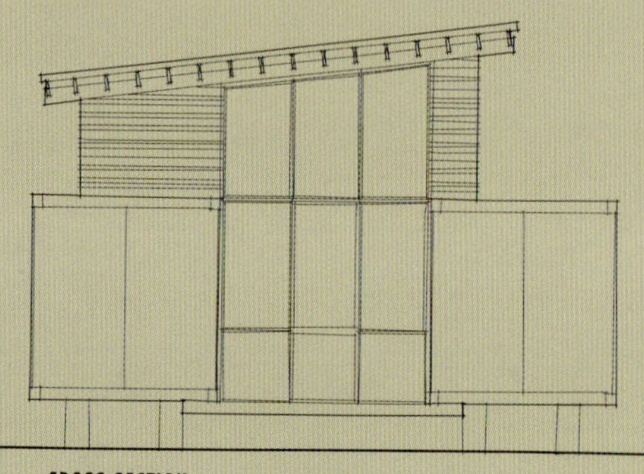

CROSS SECTION

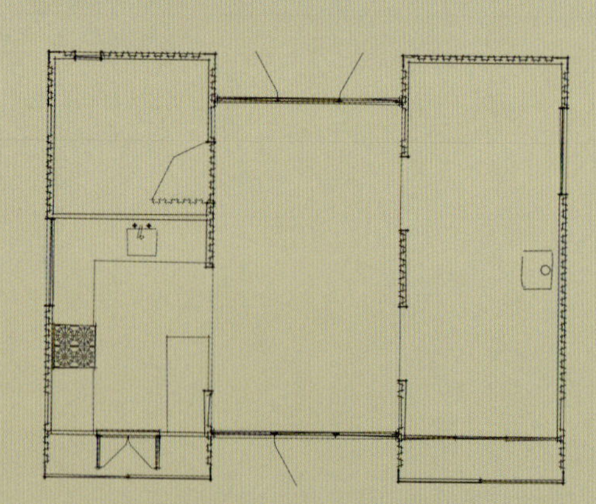

FLOOR PLAN

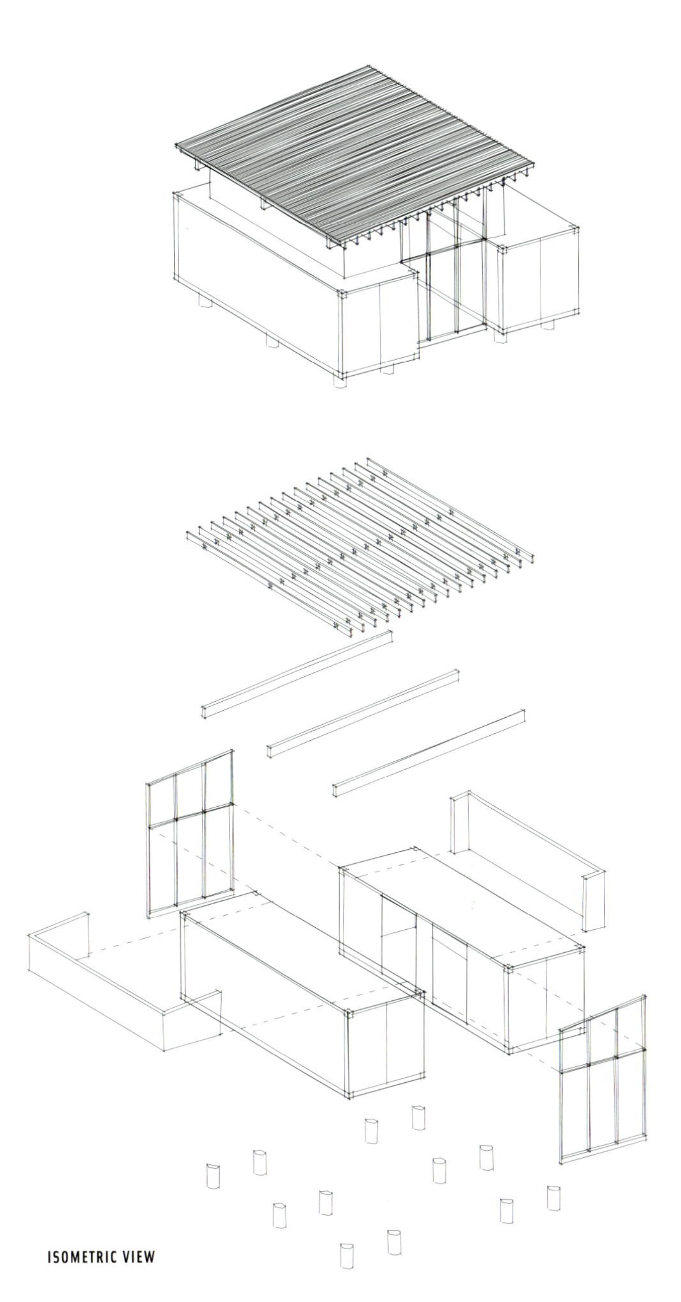

ISOMETRIC VIEW

2210
87

NAME ZIGLOO
USE FAMILY HOME ARCHITECTS KEITH DEWEY
LOCATION VICTORIA BC, CANADA YEAR 2006
CLIENT KEITH DEWEY

Keith Dewey

A small lot in a housing neighborhood of Fernwood Village, Victoria BC, is home to the Zigloo Domestique. The construction of this one-family container house was given extensive on-line coverage, therefore offering all fans of self-constructed container housing an excellent insight into the erection of a container home.

Architect Keith Dewey based the Zigloo Domestique on three R's: making use of run-down shipping containers (Recycle, Reuse), he employed little other material (Reduce) and constructed the Zigloo Domestique with only one quarter of the quantity of timber that usually makes up an average Canadian wooden frame house.

The Zigloo Domestique consists of eight shipping containers, stacked into three levels of homey prefab space. Containers rest on concrete floors that top off a basement housing the laundry room, bathroom and recreation room. The elevated ground floor contains the living room, kitchen and dining room, while another bathroom and two bedrooms make up the first floor.

As opposed to some container projects, which tend to showcase the container origin of the building on the outside as well as on the inside, the Zigloo Domestique has disguised its interior with drywall so as to make it look like a classically built house. The exterior, on the other hand, which is painted in an industrial enamel high gloss paint that is typical of shipping containers in service today, does not hide its origin. On the contrary, the front wall boasts the marking of the freight container it is made of. The curved roofs covering container units help to soften the angular nature of the construction.

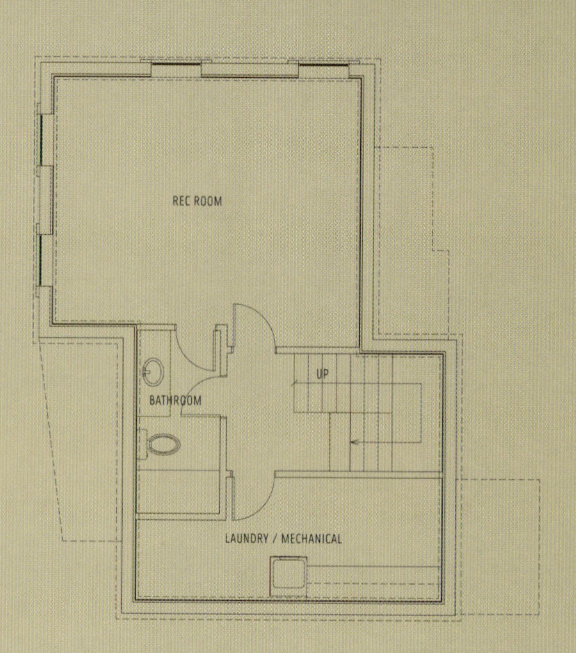

BASEMENT FLOOR PLAN

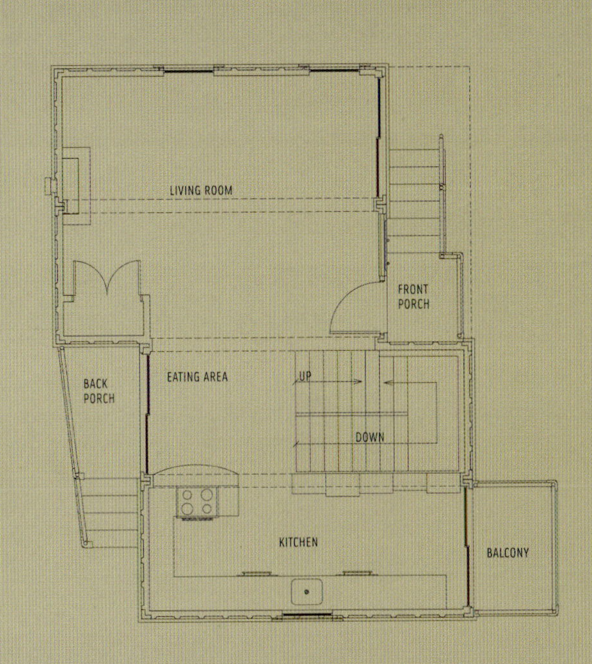

GROUND FLOOR PLAN

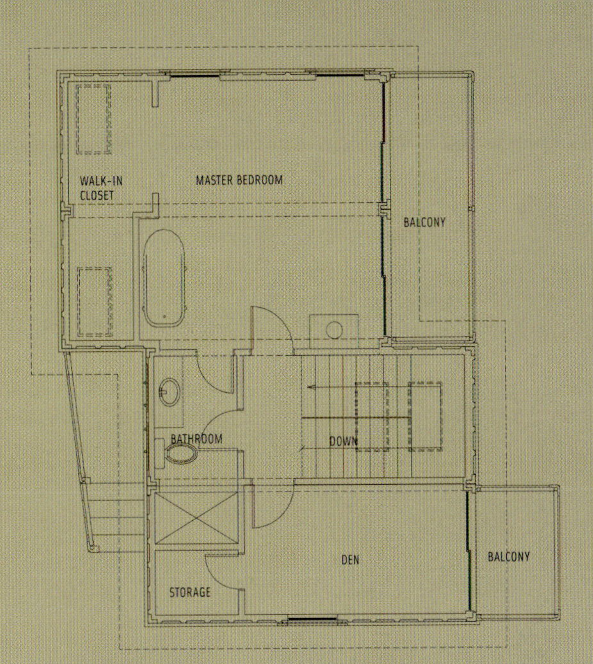

FIRST FLOOR PLAN

ECMU 103501 4
FR 2210
NET
CU CAP
NET
CU CAP

The story of this house begins when Ross Stevens, industrial designer and Professor at Victoria University in Wellington, New Zealand, decides to set up a simple residence. He is not looking for a home or a place to get emotionally attached to, but merely a place where he can stay during the week while teaching at university.

The building developed gradually, with the designer being true to his philosophy of trying to build with components that were currently available; what was important to him was how objects looked, not what they in fact were. Stevens chose a site in a Wellington suburb – an abandoned hole near the local rubbish dump – and built his house using refrigerated containers, tower crane sections, fire escape stairs, and industrial waste steel.

The house, which is located in a narrow opening of a rocky slope, rises upwards from the garage on ground level, which rests on a steel construction. Since three sides of the narrow opening are pure rock, with the fourth being the road below, the three dark-grey 40' ISO containers above the garage had to be placed one on top of the other. A steel staircase leads up from garage level to the deck on the first floor, where the house has its entrance and inside a kitchen, living room and toilet. An external staircase leads further up to the roofed deck on the second floor, which the inhabitants use as a greenhouse. The second floor has two bedrooms and a bathroom, and is connected with the third container by an in-house spiral staircase, making the upper two storeys of the house a duplex. The third, topmost floor has no partitions and contains a sleeping-living area and a small balcony. The furniture is retro, varied in style, and seems to have been given – much like the shell-construction elements – somewhat of a second chance. The front facade has no openings but a window on the top floor and 20 optical lenses, which makes the container house shielded by solid rock appear like a super modern fortress. The back side of the house, conversely, has many openings, making the rock visually blend in with the building and become part of it.

NAME ➔ **STEVENS CONTAINER HOUSE**
USE ➔ SECOND HOME ARCHITECTS ➔ ROSS STEVENS
LOCATION ➔ WELLINGTON, NEW ZEALAND YEAR ➔ 2006
CLIENT ➔ ROSS STEVENS

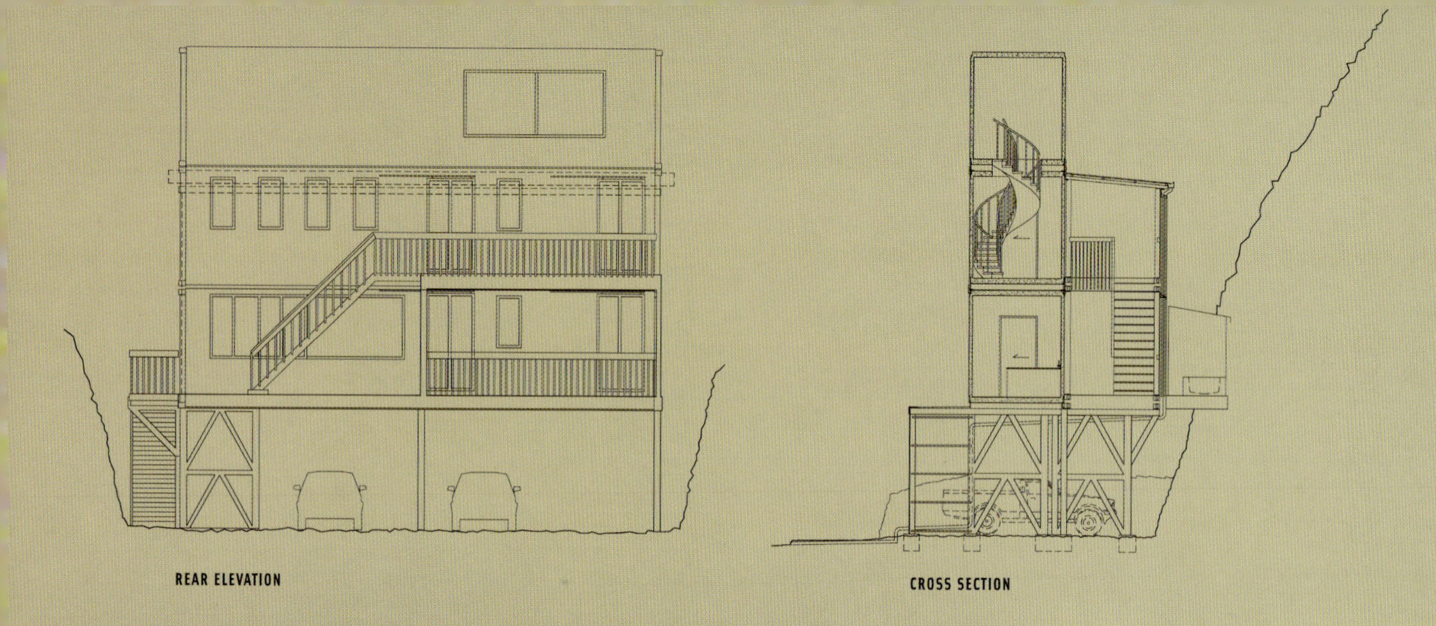
REAR ELEVATION
CROSS SECTION

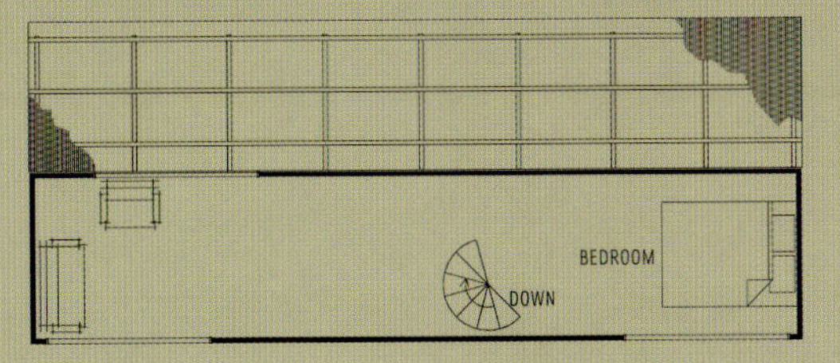

LEVEL THREE

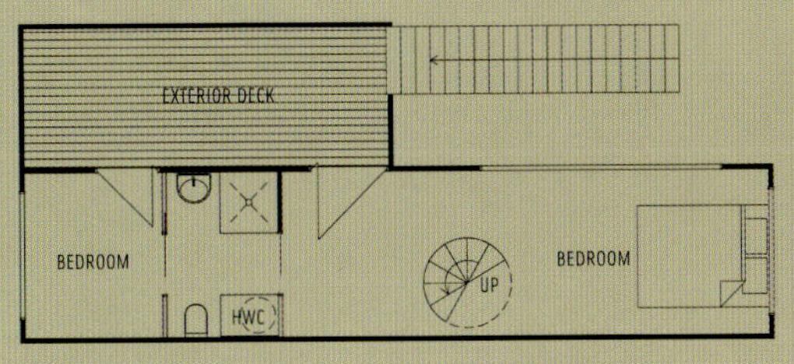

LEVEL TWO

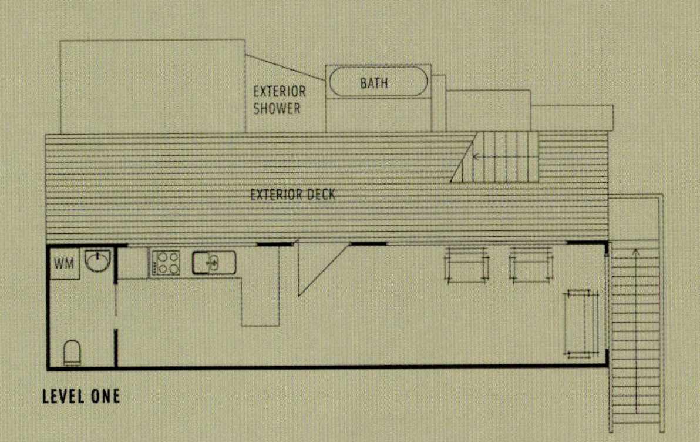

LEVEL ONE

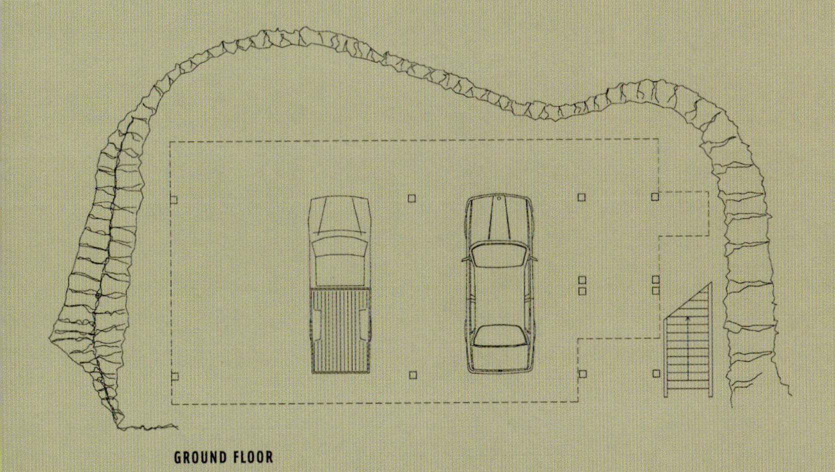
GROUND FLOOR

MOLU 516 0687
JP 4532

NAME CHALET DU CHEMIN BROCHU

USE ONE-FAMILY HOUSE ARCHITECTS PIERRE MORENCY ARCHITECTE LOCATION BEAULAC-GARTHBY, QUEBEC, CANADA YEAR 2006 CLIENT MORENCY FAMILY

Piere Morency Architecte

A stone's throw away from Lake Aylmer in Canada's Chaudière-Appalaches region, architect Pierre Morency set up a second home for his family. On account of the beautiful natural setting, an important aspect was environmental protection. There was no need to cut down adjacent trees for construction purposes, and nature-friendly materials were used. The main body of the house is three recycled shipping containers painted black, the interior of which is covered in recycled wood. The entire family participated in planning the building, and especially the children's wishes were taken very seriously; one of the sons wanted a spacecraft, another wanted a tree house, and both had their wish come true in a way. The top black container box floats away from its base, as if in space: the spacecraft. The building's position among the trees, the outside wooden paneling and elevated first floor give it the air of a tree house. The construction line runs from a small cape on lake shore and cuts right through the center of the house, where the two ground-level containers stand apart by the width of the wooden staircase.

The chalet has three layers. The exterior is covered in wooden paneling, underneath are the black recycled steel containers, while the interior has another wooden coating. Containers rest on a concrete base, which descends below ground level and hosts a guest suite and service room. The ground floor of the house consists of two containers standing apart and a central wooden area connecting them, and accommodates spaces such as the kitchen, dining room and living room. Further up is the third container, rotated to the side and supported by pillars. Here the atmosphere is more private, with the container hosting bedrooms for the children, bathroom and master bedroom. The master bedroom lies in the wooden box stretching out of the container, offering breathtaking views of the lake. The shorter sides are all glass and there are also many other windows, so that the house has plenty of daylight and the family gets a heightened feeling of truly coexisting with the surrounding nature.

SITE PLAN

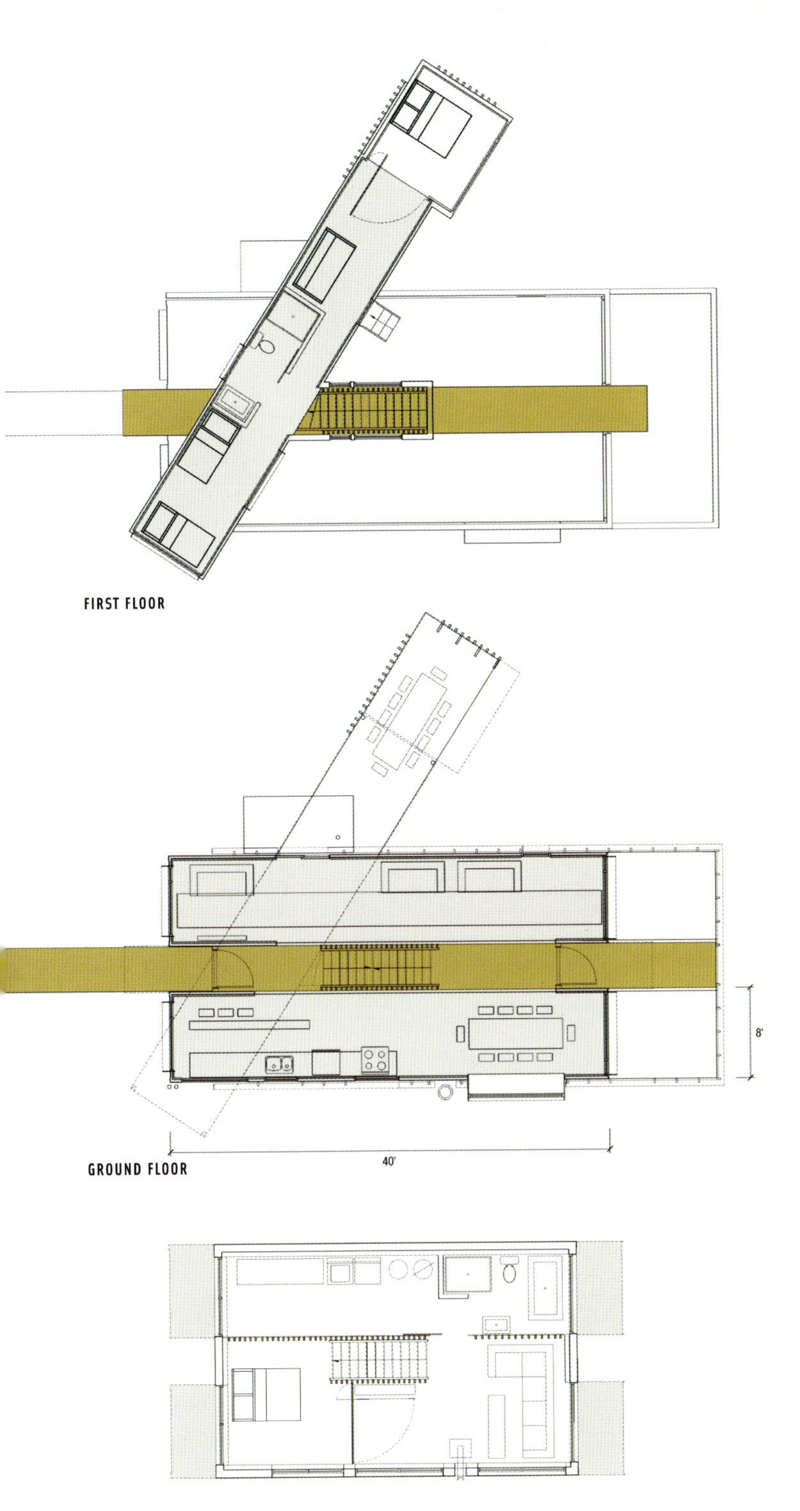
FIRST FLOOR
8'
40'
GROUND FLOOR
BASEMENT FLOOR

NAME REDONDO BEACH HOUSE
USE FAMILY HOUSE ARCHITECTS DEMARIA DESIGN ASSOCIATES LOCATION REDONDO BEACH, CA, USA
YEAR 2007 CLIENT PRIVATE

This eye-catching single family residence on California's Redondo Beach is a container-based building that also employs a combination of conventional stick frame construction and prefabricated assemblies. The building's core consists of recycled steel shipping containers, chosen as building blocks to control project time and cost without compromising design quality. In a part of the country vulnerable to earthquakes, the structure's sturdiness has clear appeal, while containers are also mold-, fire-, weather- and termite-proof. Aside for containers, this custom-made house is special for its use of construction methods and materials untypical in traditional housing techniques.

Eight containers of various sizes, all painted white, are placed onto a concrete base. They are stacked in two levels, some perpendicular to the others, and the spaces created in between them are framed in wood and steel. This very interplay of different construction elements creates unusual spatial situations and animates the house. The four larger containers are divided into utility spaces, such as the library, laundry, bathrooms and large wardrobes, while the smaller containers house bedrooms and the kitchen. There is an additional, invisible container in the back garden, sunk into the ground and filled with water – a swimming pool. The largest room in the house is a double-height living room, with a fireplace in the center, around and above which the client would like to set up a recreational climbing wall. The living room has two corner walls made of airplane-hangar doors that open up upwards to unite the house with the back garden, and serve as awnings when in the open position.

This beach house displays its container aesthetics on the out- as well as inside; there is no interior or exterior sheathing, since the walls have been covered in a special ceramic-based insulation coating, the kind used on NASA's space shuttle. The construction incorporates several environmentally sensitive and energy-saving elements, such as formaldehyde-free plywood, greenhouse acrylic panels, denim insulation, natural ventilation instead of air conditioning, etc. Being technologically advanced, prefabricated and highly customized, this environmentally conscious and affordable "hybrid" home is fully equipped to meet the needs of present and future home-seekers.

1. LIBRARY/GUEST
2. LAUNDRY
3. BALCONY
4. BATH
5. BEDROOM
6. HALLWAY
7. HIS CLOSET
8. MASTER BEDROOM
9. HER CLOSET
10. MASTER BATH

FIRST FLOOR PLAN

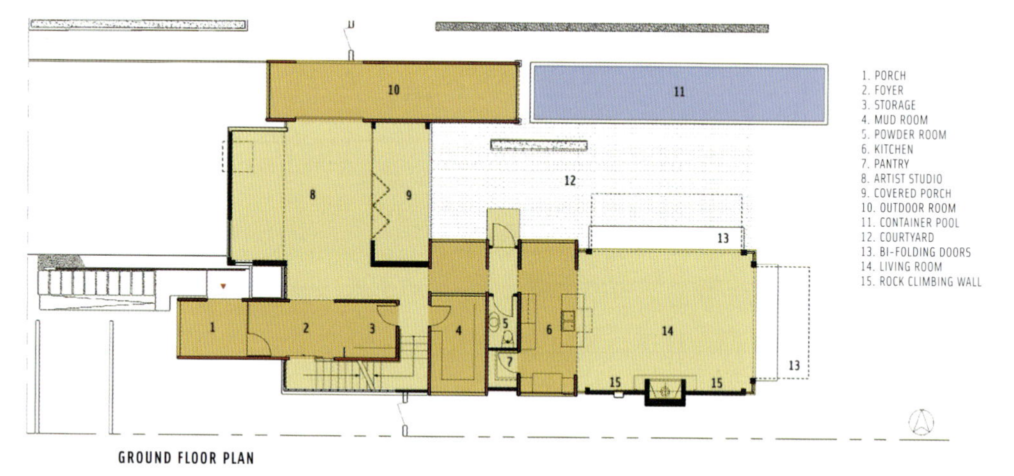

1. PORCH
2. FOYER
3. STORAGE
4. MUD ROOM
5. POWDER ROOM
6. KITCHEN
7. PANTRY
8. ARTIST STUDIO
9. COVERED PORCH
10. OUTDOOR ROOM
11. CONTAINER POOL
12. COURTYARD
13. BI-FOLDING DOORS
14. LIVING ROOM
15. ROCK CLIMBING WALL

GROUND FLOOR PLAN

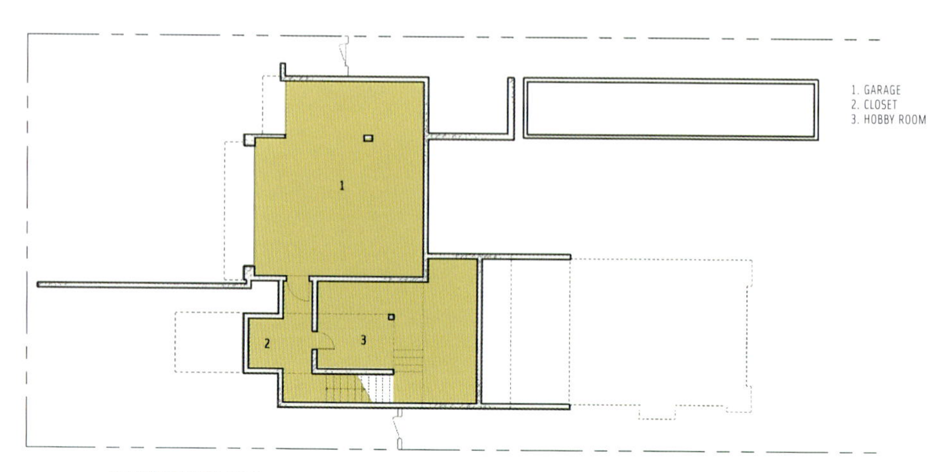

1. GARAGE
2. CLOSET
3. HOBBY ROOM

BASEMENT FLOOR PLAN

12 CONTAINERS

NAME→ **12 CONTAINER HOUSE**

USE→ VACATION HOME ARCHITECTS→ ADAM KALKIN

LOCATION→ BROOKLIN, MAINE, USA YEAR→ 2002

CLIENT→ ADRIANCE FAMILY

The 12 Container House is a custom-made prefabricated summer home, which architect Adam Kalkin created for the Adriance family from 12 recycled shipping containers. In return for their confidence in his choosing containers for the job, Adam Kalkin has surely designed for his clients a spacious residence of quite an extraordinary appearance. Constructed back in 2002, it takes up some 4,000 sq feet (372 m²) and remains to this day probably the largest container residence ever. It is situated on a private plot outside the city of Brooklin in Maine, surrounded by trees and only footsteps away from the sea shore. It attracted a lot of attention during construction, especially from the locals. This is only natural, since it is not every day that you get a container dwelling join your neighborhood, let alone one of these proportions.

Twelve orange containers are placed on a concrete base. Stacked into two levels they make two letters T, which are at a distance to each other and have a large glazed area between them. This central two-level space lies at the heart of the symmetrical house and hosts a spacious living area, dining room and two staircases, which lead onto the upper floor, each into its own T-winged part of the house. Containers on the ground floor house the kitchen, library, office, playground and guest bedroom. Abundant light penetrates the house through floor-to-ceiling windows and all the containers' shorter sides, which are all glass on the outside as well on the on the inside. This allows additional light to enter from the house's central space, and enables visual communication between the dwellers inside the house. The containers' longer sides have no openings on the outside, but have been fully cut out on the inside in some parts (like the kitchen or the library) to allow free passage among the surrounding rooms and the central space. The upper floor is more intimate, housing bathrooms and bedrooms and an office. The side of the house facing the sea has a large terrace with an outdoor fireplace.

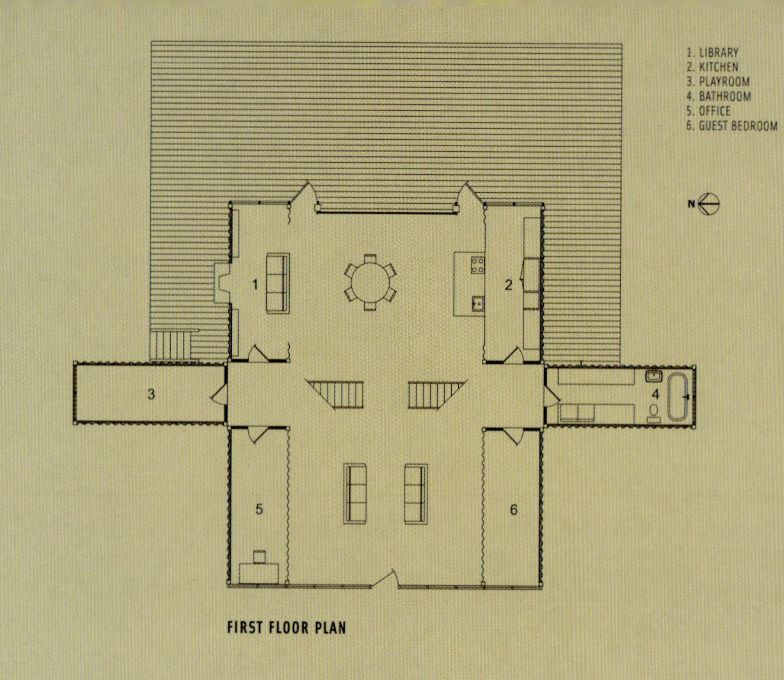

FIRST FLOOR PLAN

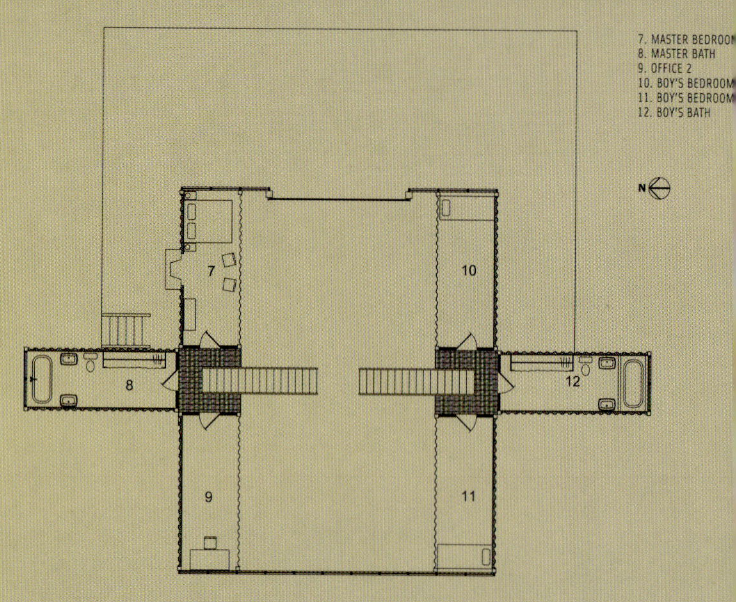

SECOND FLOOR PLAN

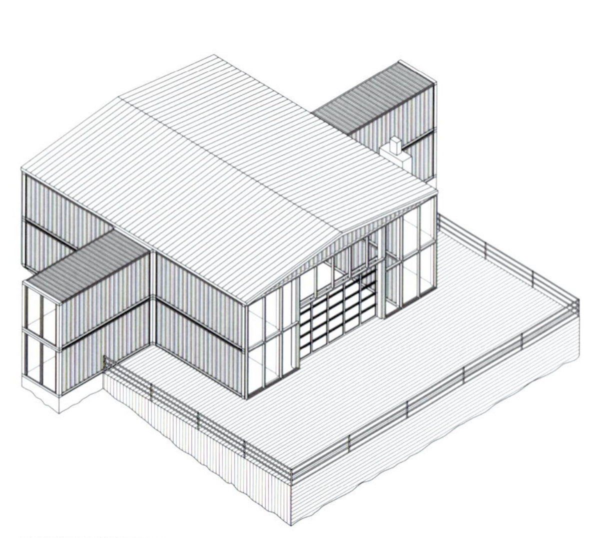

SOUTHEAST ISOMETRIC VIEW

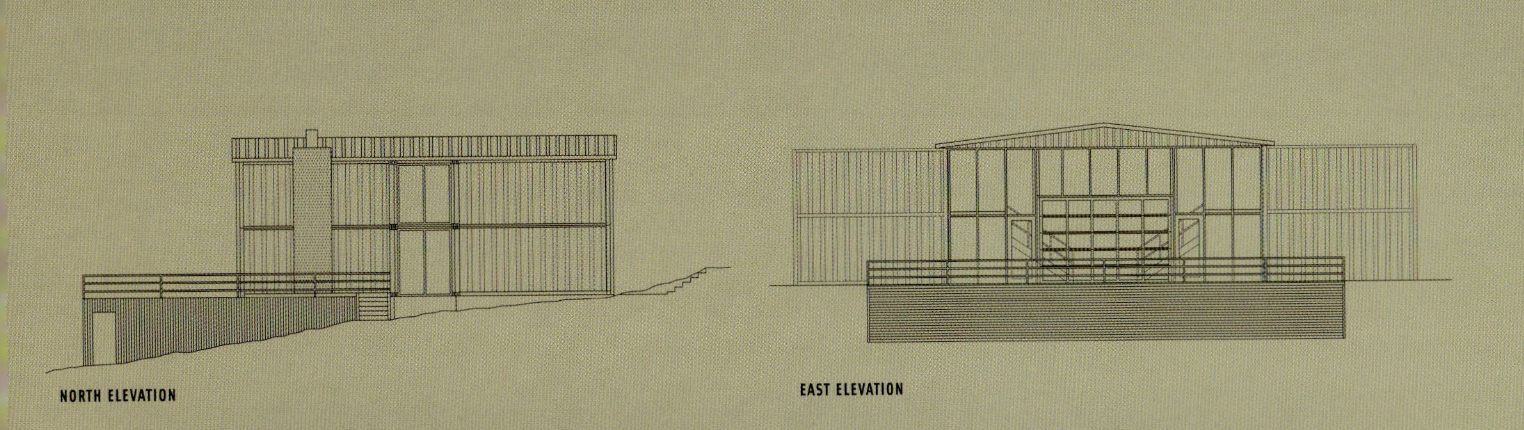
NORTH ELEVATION
EAST ELEVATION

NAME CONTAINER CITY I

USE LIVE/WORK AND STUDIO SPACE ARCHITECTS NICHOLAS LACEY AND PARTNERS LOCATION LONDON, ENGLAND, UK YEAR 2001 CLIENT URBAN SPACE MANAGEMENT LTD

The Trinity Buoy Wharf lies in the eastern part of the Docklands, opposite the Millennium Dome, where the River Lea flows into London´s Thames. Urban Space Management won a competition to develop the site into a cultural centre, accommodating live/work studios, and serving as a venue for exhibitions and other events. Container City I was one of the first realized container projects in the world that received wide public coverage.

It is sited in the open space between existing warehouses and made up of 20 recycled 40' containers. Initially they were stacked in the classical manner into three levels, but due to high demand a fourth floor was added, so that now Container City I has 15 live/work apartments. They are sublet in the long- or short-term and mostly inhabited by artists and people of other creative occupations. Once the containers were assembled, the partitions between them were removed and the interior rooms separated arbitrarily. Each storey therefore comprises several studios of different sizes. The entrance takes the form of the external staircase, fitted into a vertical 40' container, which connects to the building through short bridges on every level.

Containers are all russet, to make them blend in with the brickwork of the local industrial buildings. The typical outside pattern is created by round windows and balconies wedged between the two door wings of standard containers. Daylight is admitted into the studios through round windows of two sizes, and through the glazed shorter side of the container, where sliding doors open onto the balcony. Toilets are located at the bridge entrances, individual units have but minimum services, and electric heating is used.

As well as being environmentally friendly (with over 80% of the building created from recycled material), Container City I is very cost effective, which is why it has already received a sequel near-by – Container City II.

TYPICAL FLOOR PLAN

30 CONTAINERS

NAME CONTAINER CITY II

USE LIVE/WORK AND STUDIO SPACE ARCHITECTS NICHOLAS LACEY AND PARTNERS LOCATION LONDON, ENGLAND, UK YEAR 2002 CLIENT URBAN SPACE MANAGEMENT LTD

Container City II is both an extension and evolution of the first building, with which it is connected on the Trinity Buoy Wharf by means of a joint central staircase, fitted inside two vertically placed 40' containers. A system of inter-connecting bridges runs from the staircase into both buildings, with the topmost bridge being covered in a fabric tension roof. Featuring a new lift, both buildings are also fully accessible to the disabled.

Over five levels, Container City II accommodates 22 studios. All rooms are covered in spray insulation over membrane waterproofing to stop the condensation, and the walls and ceiling are finished in plasterboard. Installation blocks have been minimized, and electric heating is used.

Like its peer, Container City II has typical round windows and balconies, which here also appear on the longer sides of containers. The chief difference between the two buildings is their color and how containers are stacked. Here, some are perpendicular to the others, which results in a large, complex arrangement and overhanging container boxes, some of which rest on secondary steel framing, to make space for vehicles passing beneath the building. Containers are painted in alternating different bright colors (yellow, orange, red and white) to make each individual unit stand out and reflect the creative flair of those who work there. Funky shapes and colors, however, are not reflected in the interior, which is – surprisingly perhaps – quite ordinary. This very discrepancy between the building's outside and inside helped convince the skeptics who reject container housing claiming it is too futuristic or minimalistic in its preoccupation with design. Container City II is evidence to the fact that an ordinary apartment is indeed possible inside a container shell, which makes container housing one step closer to the general public and explains the huge success of the project.

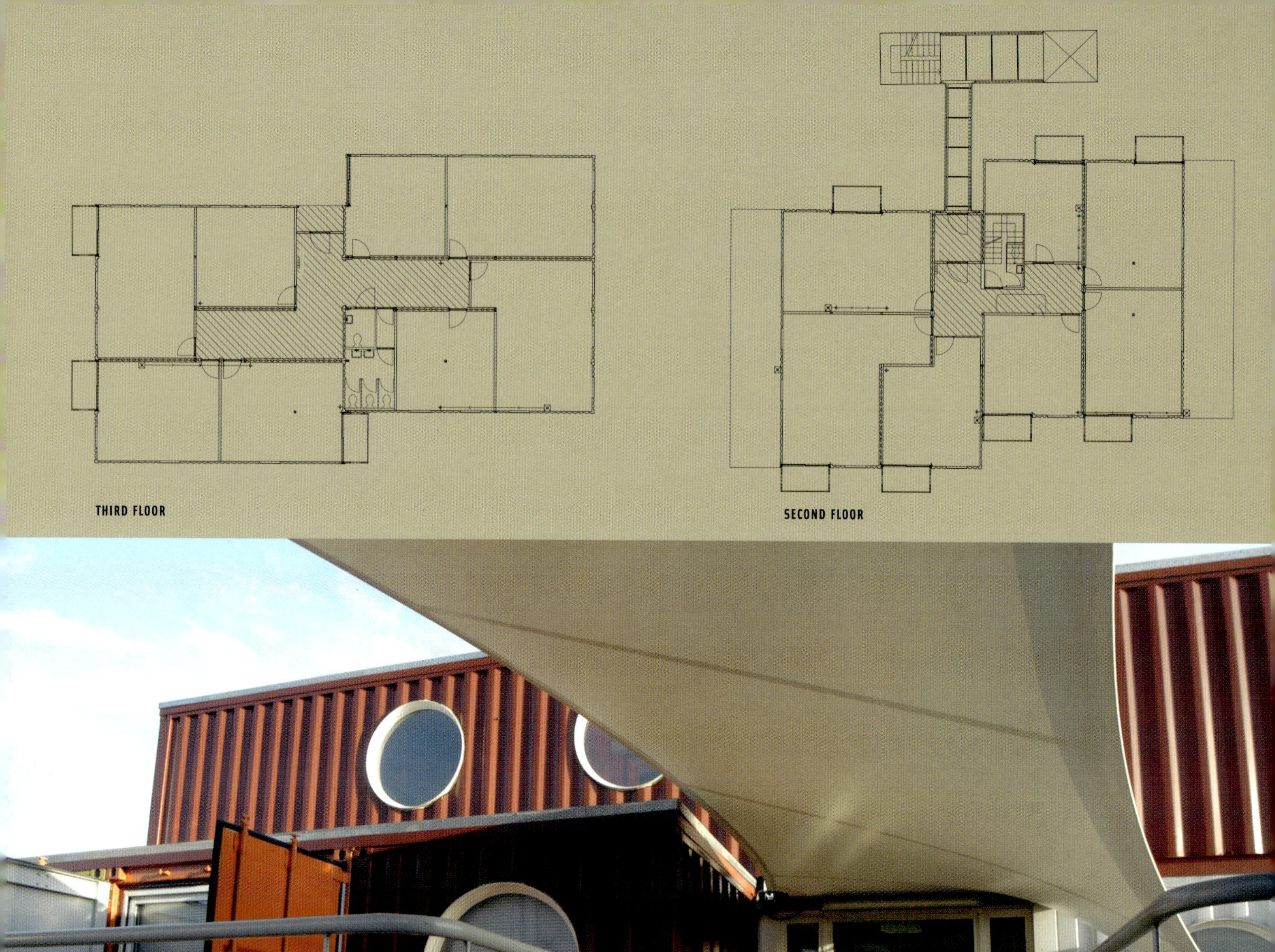
THIRD FLOOR
SECOND FLOOR

NAME THE RIVERSIDE BUILDING
USE STUDIO/OFFICE ARCHITECTS ABK ARCHITECTS
LOCATION LONDON, ENGLAND, UK YEAR 2005 CLIENT URBAN SPACE MANAGEMENT LTD

The Riverside Building is the third Container CityTM project. Situated on the banks of the river Thames opposite the Millennium Dome, in a rapidly developing London business district, it is adjacent to the Container City buildings. All such container projects offer ever new visual and technical solutions, which goes to show that in container architecture, monotony is not to be feared.

The Riverside Building comprises 73 recycled containers set over five floors, and hosts 22 office spaces of various sizes. The entire structure took only eight days to erect. The ground floor has a bar and cafe/restaurant facing the Thames, while all the other spaces are studios, offices and apartments. The smallest unit comprises two containers and comes to 27m² (291 sq ft), with the largest being composed of ten containers and offering 141m² (1518 sq ft) of room. The interior walls between containers have been removed so as to create a large open layout and allow for a custom-made organization of the interior. The rooms get plenty of daylight, since parts of the facade are glazed and windows have been added on the sides. Details of the structure comply with the current trends and move away from typical container particularities.

The building has been designed with containers running around an atrium and artificial pond in the centre, where two upright containers host the central staircase and elevator. From there, external decked walkways and balconies lead to individual studios, while the building itself may also be accessed over water, being in the direct vicinity of the pier. The Riverside Building takes full advantage of its riverside location, providing spectacular views at an affordable price in what is developing into a prosperous London business area.

ABK Arhitects have designed an airy, modern and affordable building, constructed in a large part from recycled materials, yet hiding its core element on the out- as well as inside, so that the building does not appear as if having a container construction at all.

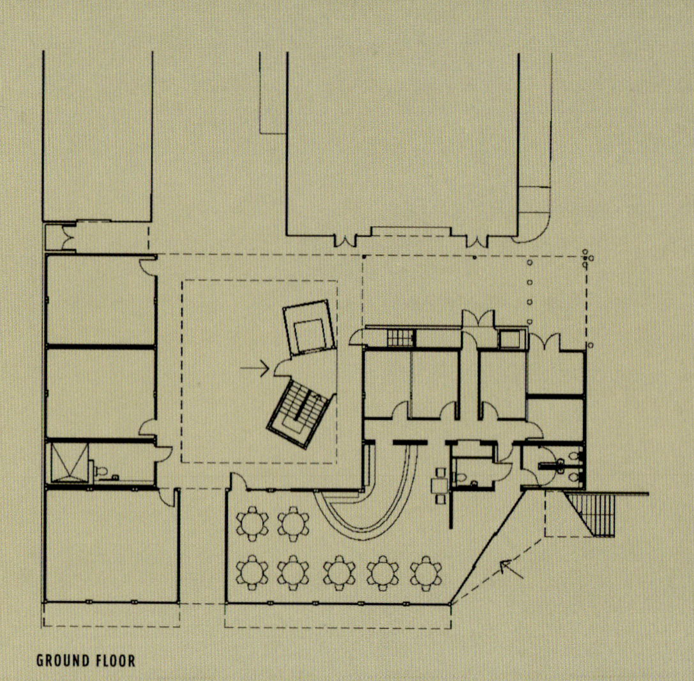
GROUND FLOOR

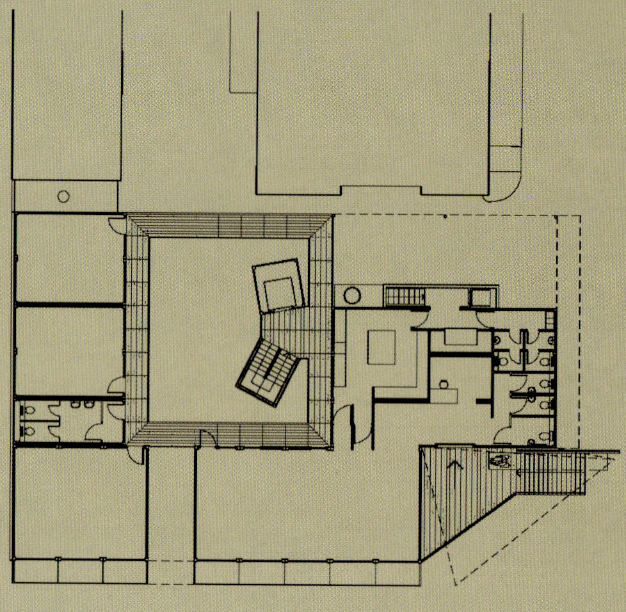
FIRST FLOOR

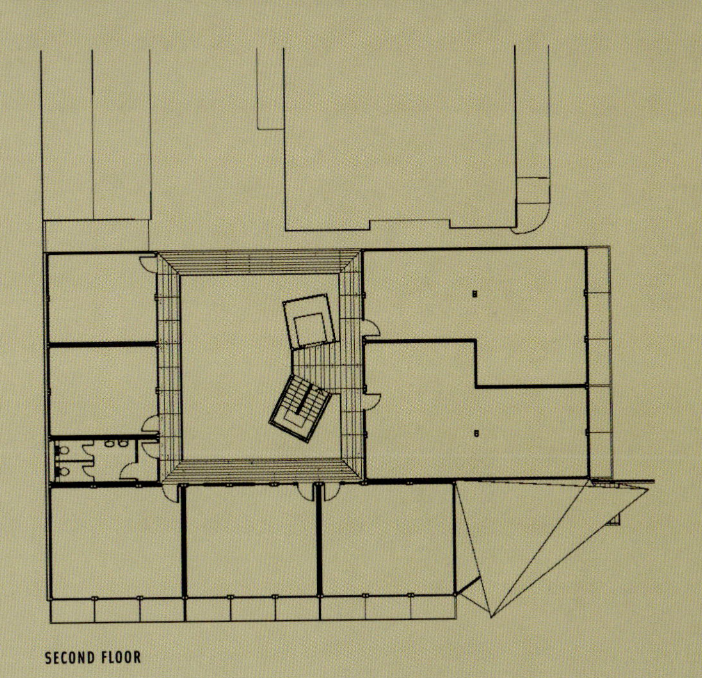

SECOND FLOOR

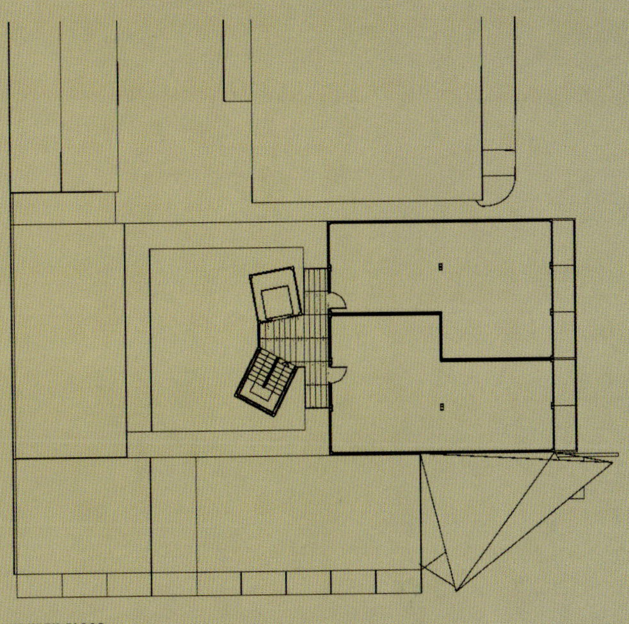

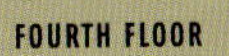

FOURTH FLOOR

24
22

1005
CONTAINERS
NAME QUBIC HOUTHAVENS
USE STUDENT HOUSING ARCHITECTS HVDN ARCHITECTEN
LOCATION AMSTERDAM, THE NETHERLANDS YEAR 2005
CLIENT WOONSTICHTING DE KEY/DE PRINCIPAAL
HVDN
Architecten
184

It is the Netherlands where container housing achieved its fullest swing in the construction of student accommodation facilities. The proverbially rational Dutch developers were convinced by containers on several points. Firstly, container constructions are quick and easy to erect and assemble, and also extremely cost-efficient. Further, a container unit can be especially tailored to the spatial needs of an individual student, while such constructions are also very flexible as to the adding of additional units when the need arises. There were several more or less successful attempts at such constructions, in the urban as well as architectural respect. Qubic, student accommodation facilities by HVDN Architecten, stands out as a particularly fine specimen.

The Qubic student accommodation facilities are located in the former dockland area of Amsterdam's Houthavens and comprise 715 student units and 72 temporary dwellings. The development also contains a ship converted into student rooms, art studios and a myriad of restaurants and bars; the lively atmosphere has made Qubic a place to be in Amsterdam.

Qubic has a strong binding form, preventing the project from looking like a container village. At the architectural level, the containers are connected by floor and roof slabs with columns in between, leaving room for attractive porches on the ground floor. The facades are constructed from storey-high modulated plastic panels that contain a variety of window openings, while colored plexiglass further adds to the lively composition of the elevations. At the urban development level, the container blocks form three courts that serve as meeting places for students; two closed courtyards with a lawn act as sheltered spaces for sport and relaxation, while the northern courtyard, facing the IJ waterfront on one side, hosts a bar/restaurant.

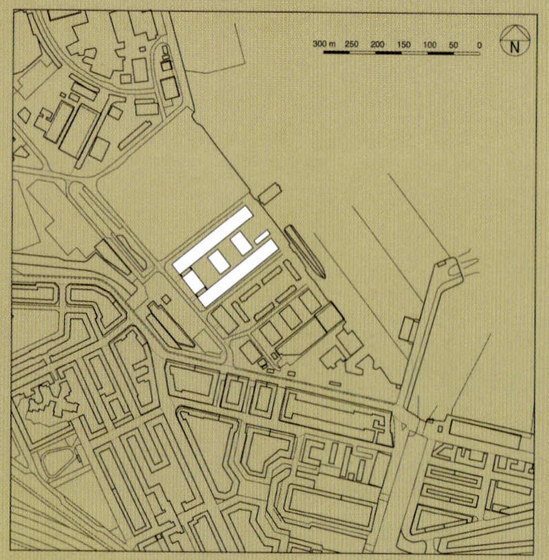

SITE PLAN

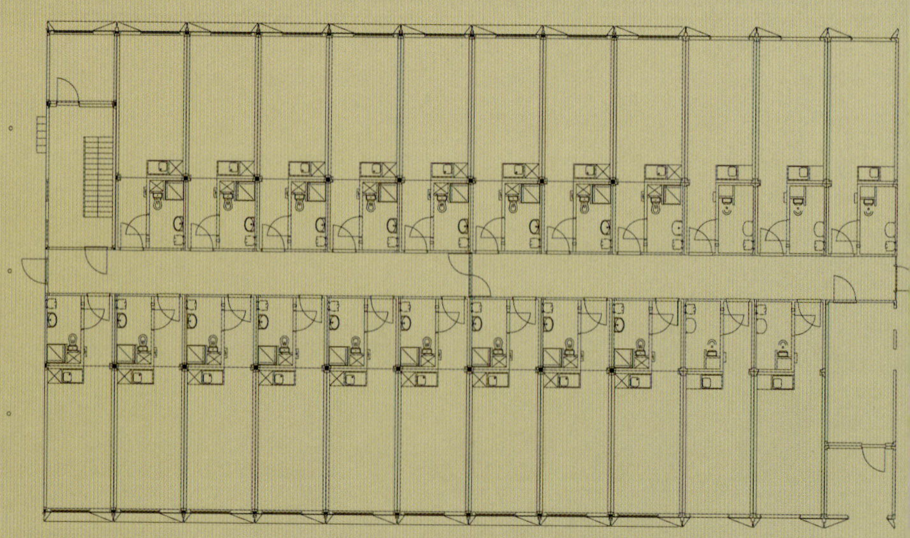

GROUND FLOOR

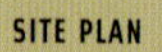

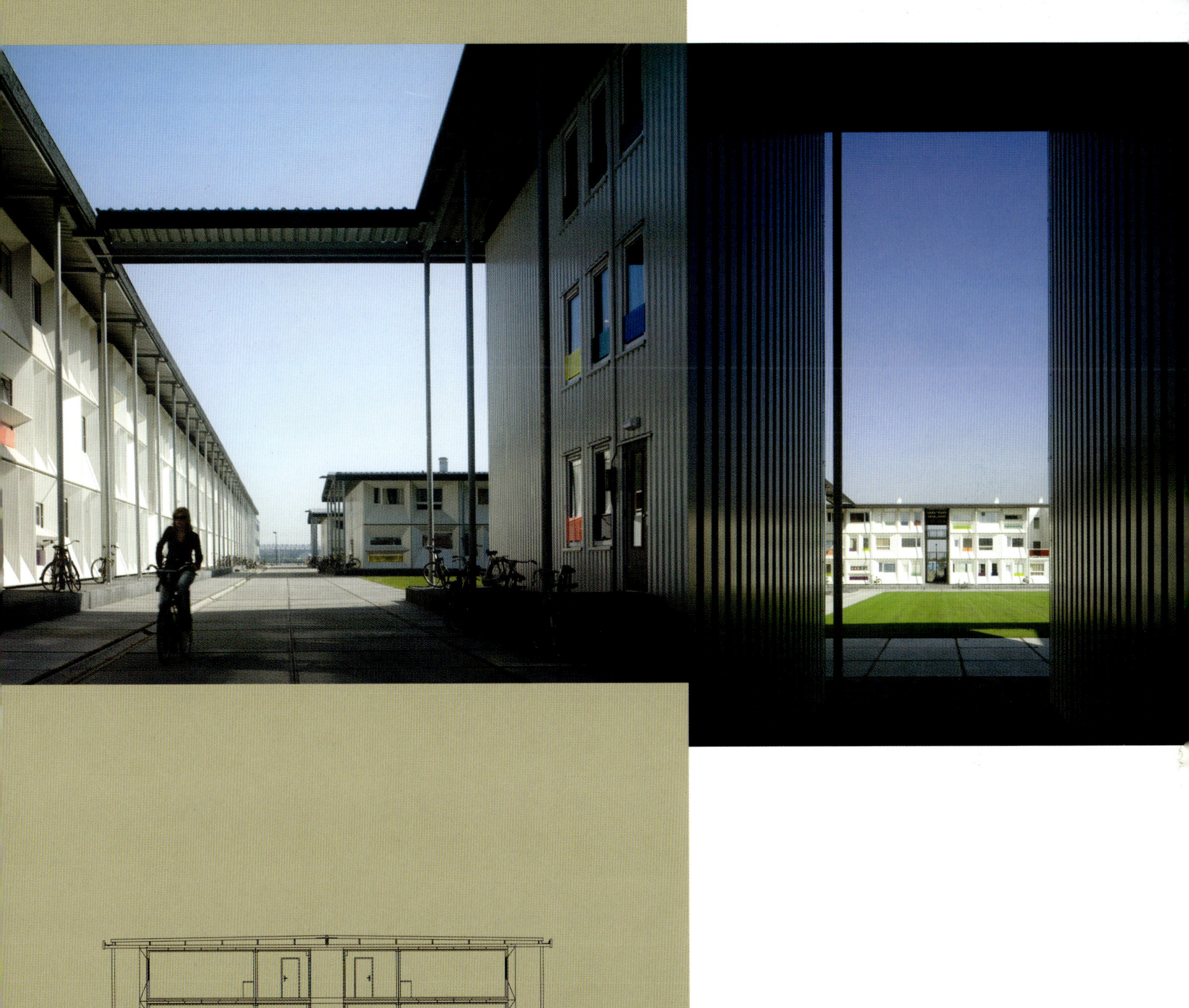

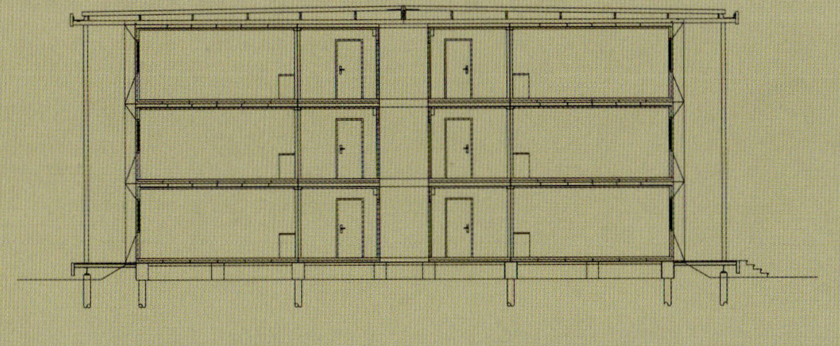

CROSS SECTION

NAME ➔ **TEMPOHOUSING/KEETWONEN**
USE ➔ STUDENT HOUSING ARCHITECTS ➔ JMW ARCHITECTEN/ TEMPOHOUSING LOCATION ➔ AMSTERDAM, THE NEDERLANDS
YEAR ➔ 2005 CLIENT ➔ WOONSTICHTING DE KEY, AMSTERDAM

JMW Architecten / Tempohousing

In a time when universities abound and differences in the quality of education they offer begin to diminish, secondary factors come into the foreground, such as student accommodation. What students need is a fully equipped, low cost residence close to their school. Such temporary student housing facilities as Keetwonen in Amsterdam can make university centres more competitive, since they successfully regulate both student bedding capacities as well as their prices.

One of the largest container campuses in the world comprises 12 blocks, united into pairs by means of joint exterior staircases that lead into individual student units. Blocks are constructed from 40' containers stacked into five levels, the top row of which is covered with a low-slope gable roof. Per block there is a services container, supplying all the units in the block with electricity and an internet connection, while the complex also includes a shop, laundry room, cafe restaurant, and a sporting area. Most of the ground space between the buildings is taken up by bicycles, the typical student means of transport, especially in the Netherlands, while there is also greenery and benches.

Each student unit is sized 30m² (323 sq ft) and is fitted inside a single container. All containers have shorter sides made of glass to allow more sunlight penetrate the room. The bathroom splits the unit into two parts joined by a passage. One part has the entrance, kitchen and dining area, while the other hosts the bedroom and living area, and a small balcony. Despite the fact that these elementary container cells are of identical composition, students can still give them character and make their temporary homes truly their own.

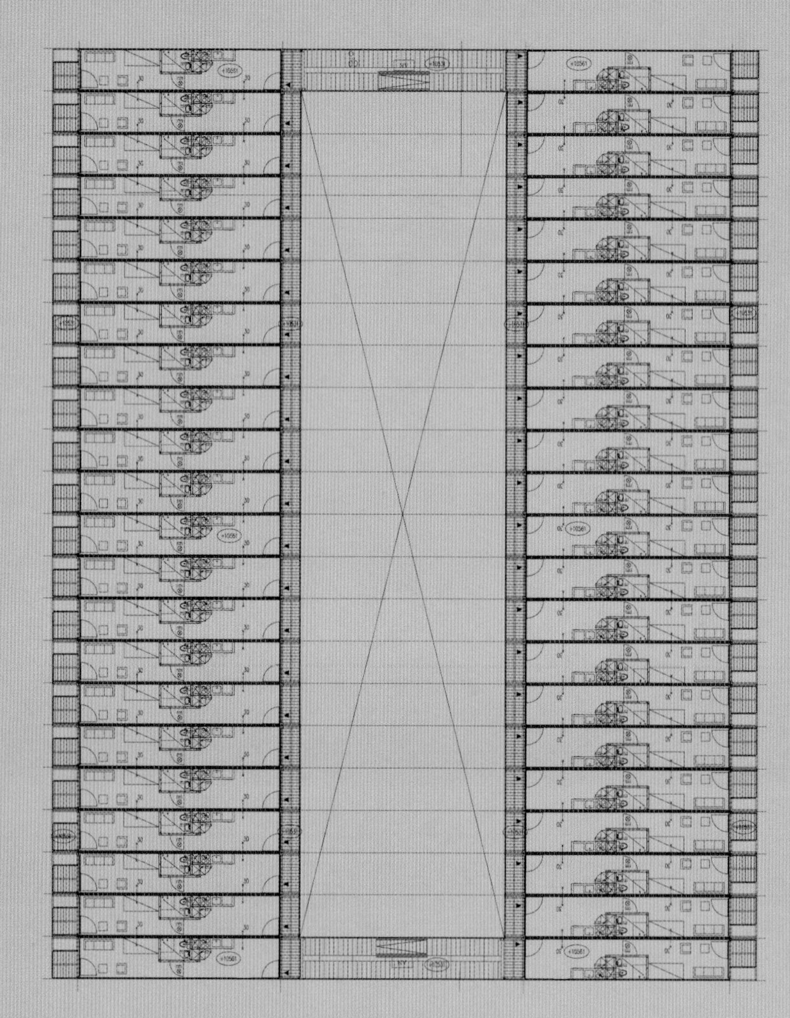

FLOOR PLAN

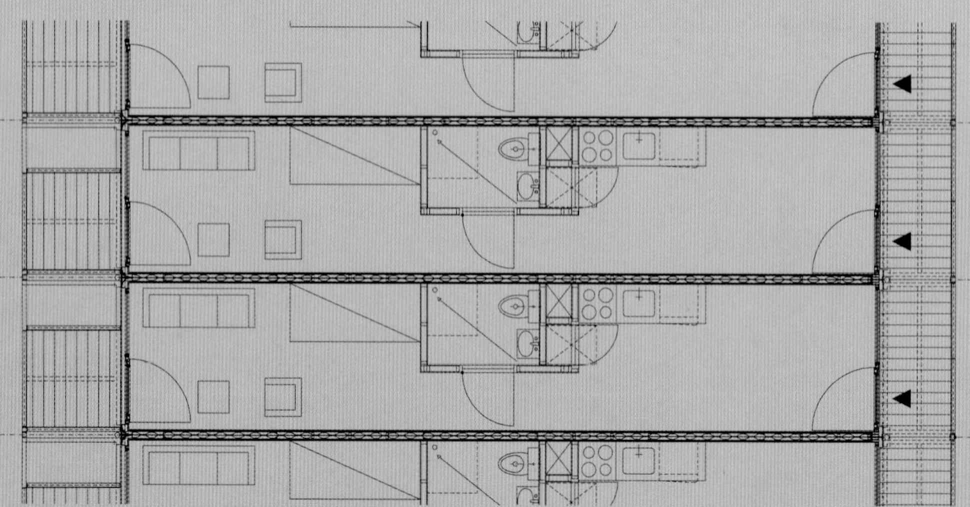

ROOM PLAN

my container is my castle
KEETWONEN IS BUILDING THE LARGEST CONTAINER VILLAGE OF EUROPE
www.keetwonen.nl
SUPERMARKT | SPORT | HORECA | WASSERETTE | KANTOREN
E

100046
100012

CHAPTER SIX EXPLAINS TWELWE CONCEPTUAL & THREE UNBUILT CONTAINER PROJECTS.

The first conceptual projects mark the period of intense development of container architecture. They include those first remodeled containers which showed that the steel box could be used for other purposes apart from transport. These constructions are usually smaller in size, and often a single container is enough to get the message across. Due to space limitations, a multi-use interior is a frequent feature. Interior design works according to the fold-in/fold-out principle, so that furniture is either neatly stored away or ready for use, depending on the need of the day. While some projects do fine with but carpenter details, others make use of more high-tech solutions driven by electricity and/or hydraulics, which can open the container up, lift it or dismantle it to increase its useful area. These expansion-oriented tendencies underline the fact that a single container unit remains spatially very limited, despite some projects wanting to show the opposite.

Pioneering conceptual projects all emphasized the cargo container's main characteristics, such as its mobility, flexibility, cosmopolitan disposition, off-grid autonomy, etc. As container architecture as such evolves, it is beginning to gravitate toward more classical and lasting structures, but conceptual projects still remain testing their limits – as any true concept should.

The realized conceptual projects in this Chapter are joined by three significant but as yet unbuilt projects that rest on a powerful conceptual framework.

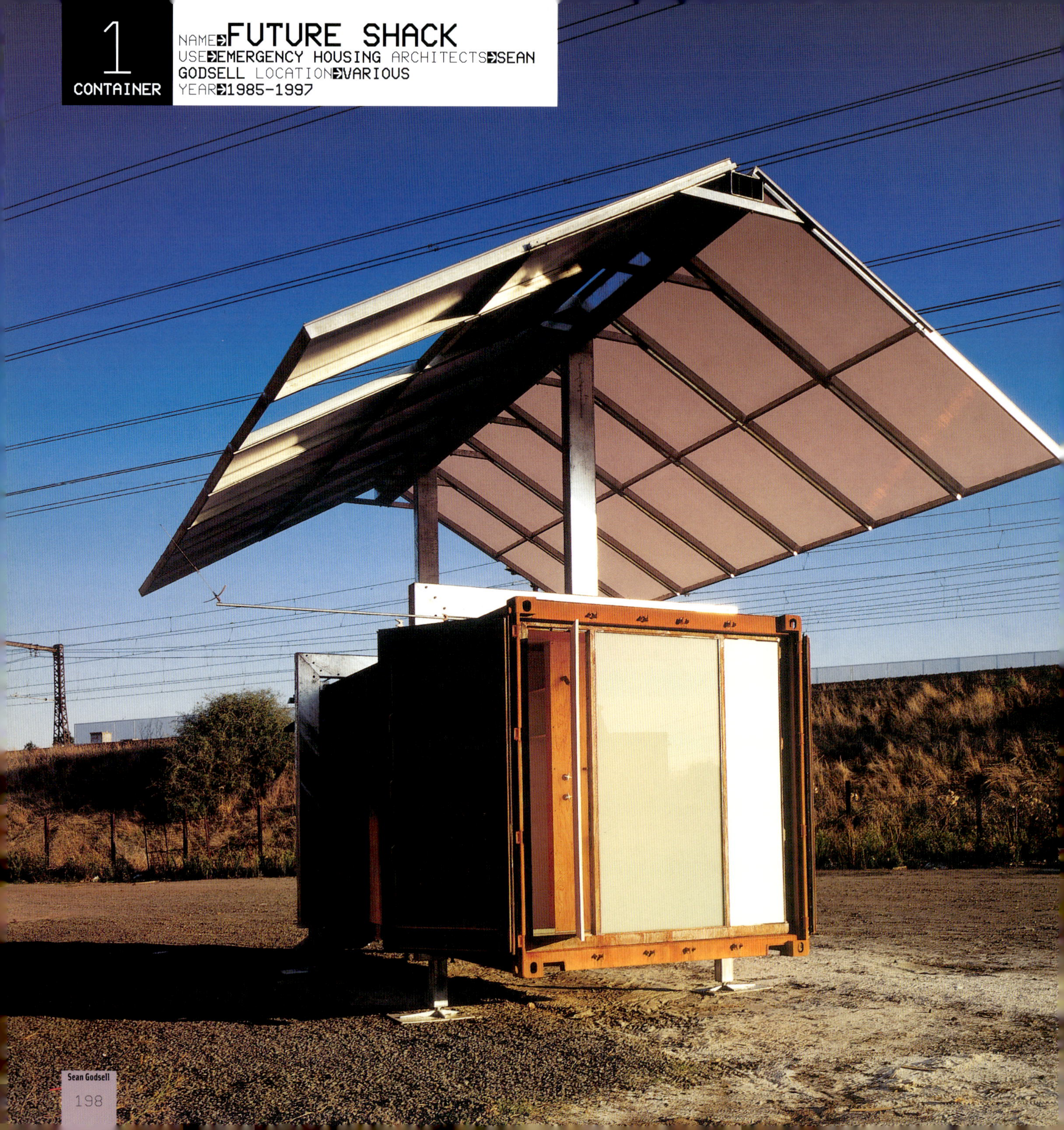

1 CONTAINER

NAME FUTURE SHACK
USE EMERGENCY HOUSING ARCHITECTS SEAN GODSELL LOCATION VARIOUS
YEAR 1985-1997

Future Shack is a prototype of mobile, mass-produced emergency and relief housing, to be used post-flood, -fire, -earthquake or similar. It could also be used to temporarily accommodate fugitives or the socially disadvantaged third-world inhabitants. The element traditionally used to transport the goods that contributed to the wealth of the developed world now assumes the function of relief for those in need. Its main advantages include its wide accessibility, low cost, sturdiness and durability, as well as the fact that it is easy to ship and can be manipulated on site.

Future Shack comprises four core parts: a recycled 20' shipping ISO container, a pair of steel brackets, a roof and access ramp. An important virtue of this container shack is its raised gable roof – a universal symbol of home, which is precisely what any raw container lacks and what the displaced people need most. When erected, the roof also provides shade, reducing the heat load on the building and creating an additional protected outdoor space. Depending on local customs, the parasol roof can be covered in indigenous materials such as thatch, mud and stick, palm leaves and so on, while it packs inside the container for transport. Two steel brackets are fixed to the outside of the container, containing four legs that telescope out and enable the module to be sited without excavation on uneven terrain, while an entrance ramp lowers to allow access to the raised container floor. The container's shorter sides are almost all glass, inviting sunlight into the room. The shack's interior is lined with plywood, and features minimalistic built-in furniture – a table and bed that fold down from the wall, a compact kitchen and bathroom, and a living area.

Future Shack has a number of environmentally sustainable design features. Photovoltaic cells can be mounted on the roof, to make the unit energy self-sufficient. The parasol roof can also be adapted to catch its own water, while the module consists chiefly of recycled materials. It can be constructed with no site works, thereby reducing site impact to the absolute minimum. When it is no longer needed, the Future Shack can be packed back into itself together with all its constituent parts and, looking like any other cargo container, relocated or stockpiled for future use.

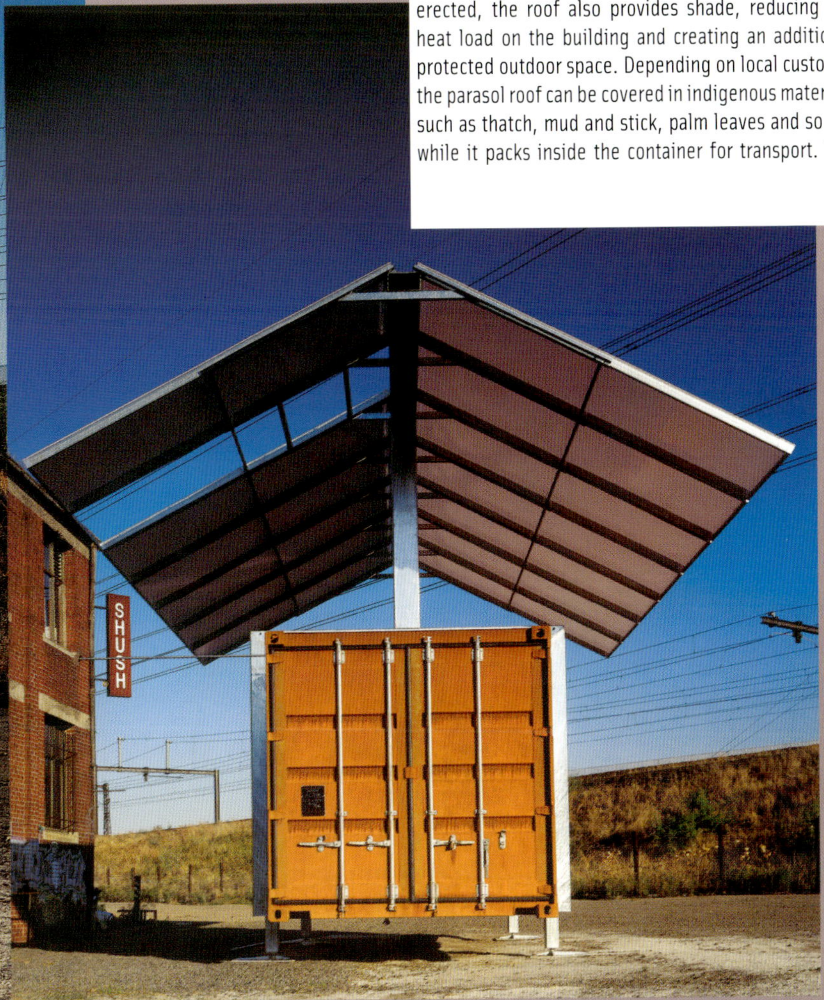

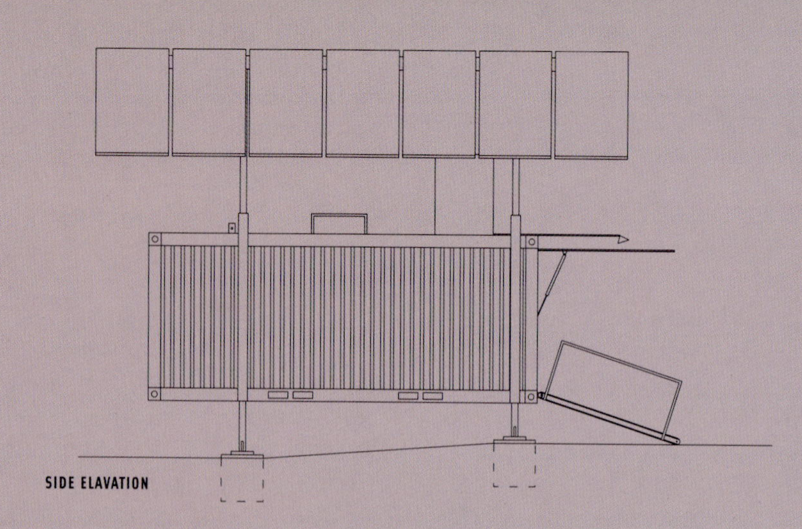
SIDE ELAVATION

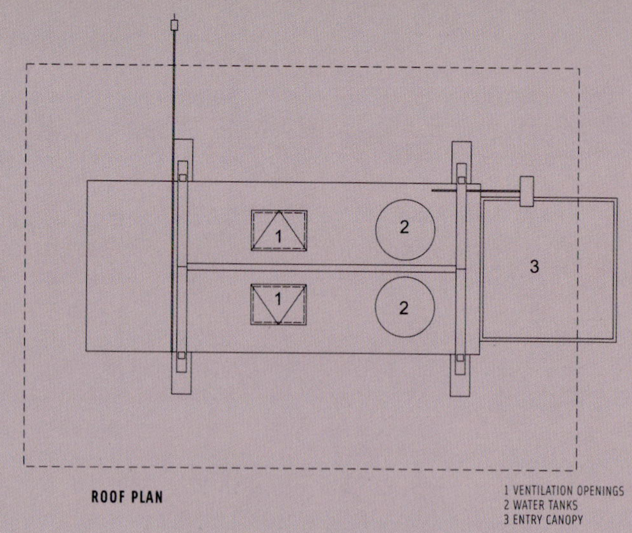
1
2
1
2
3
ROOF PLAN
1 VENTILATION OPENINGS
2 WATER TANKS
3 ENTRY CANOPY

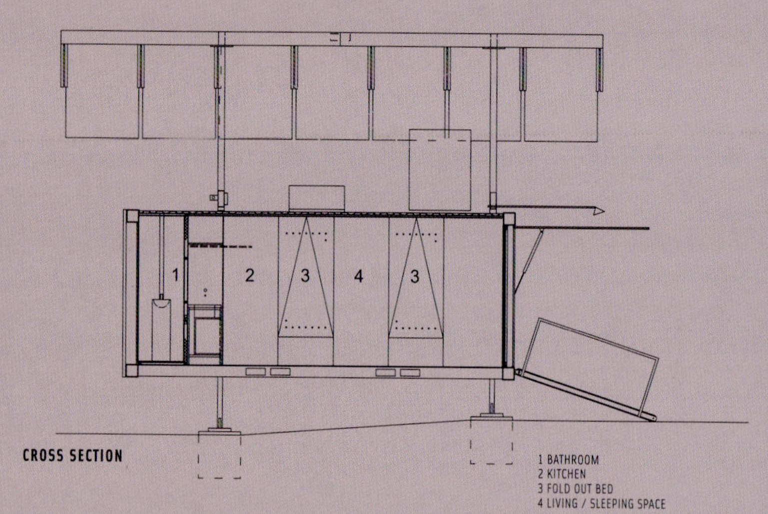
1
2
3
4
3
CROSS SECTION
1 BATHROOM
2 KITCHEN
3 FOLD OUT BED
4 LIVING / SLEEPING SPACE

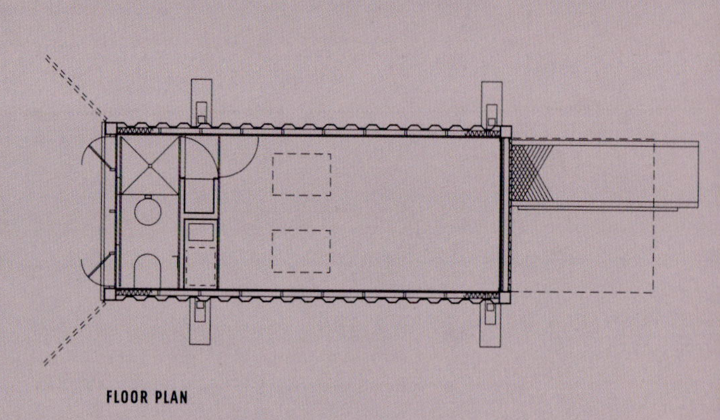
FLOOR PLAN

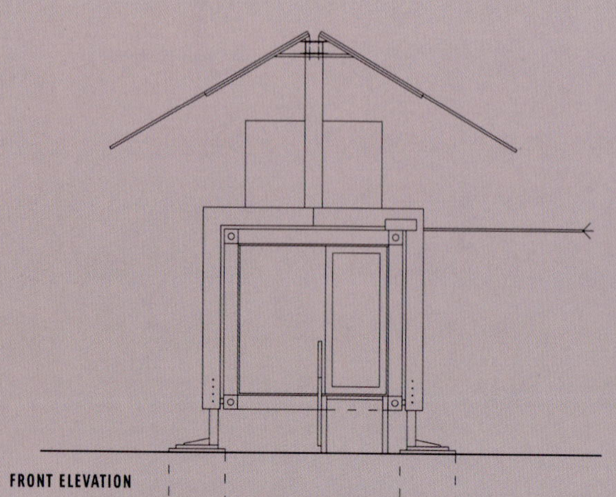
FRONT ELEVATION

1
CONTAINER

NAME PUSH BUTTON HOUSE
USE COFFEE HOUSE ARCHITECTS ADAM KALKIN
LOCATION VENICE, ITALY YEAR 2007
CLIENT ILLYCAFFE S.P.A.

The Push Button House is an apartment fitted into an ordinary 20' container, which looks just like any other cargo container on the outside. But at the touch of a button, the rusty yet perfectly solid box becomes alive and transforms into a stylish six-room apartment in a matter of 90 seconds. This is how long it takes for four motors operated by hydraulic cylinders to open up and lower the container's four sides. Each side has furniture attached to it, making up individual rooms. The contrast between the container's bleak exterior and its vivid fully-equipped starch white interior perhaps compares best to the transformation of a static cocoon into a colorful butterfly. Although with the Push Button House the process goes both ways and has endless repetitive potential. One of the longer sides has the bedroom and bathroom, and the other has the white-sofa living room. The center of the container structure is where the dining room is, while the two shorter sides open up into the kitchen and library. The useful surface of the thus-expanded house is at least threefold. It functions best in a dry climate where it can stay open, but can just as well be used inside a large closed space as an eye-catching office, bar, exhibition space, etc.

Adam Kalkin created several versions of this suitcase-packed apartment, using different equipment and different methods in which the sides of the container open up. Still, all varieties share their capacity of quick transformation into a butterfly right before the eyes of the public, which is genuinely fascinated by the sheer speed with which the device-like house folds open and by the visual contrasts it creates. It is therefore not in the least surprising that this fresh concept has been given so much attention.

The Push Button House witnessed its debut on Art Basel Miami Beach in 2005, where it was a part of the exhibition, but soon a renowned Italy-based coffee maker found a new use for it. At the Venice Biennale of Art, the Push Button House became a coffee shop, in which guests to the biennale could rest and relax with a cup of delicious coffee. The most recent version of the Push Button House is again a coffee house, which was set up at the end of 2007 in New York's Time Warner center and which opens up independently of the geometry of its sides.

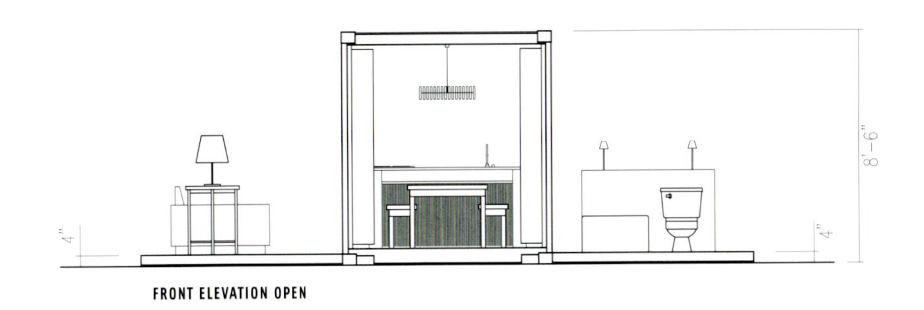

FRONT ELEVATION OPEN

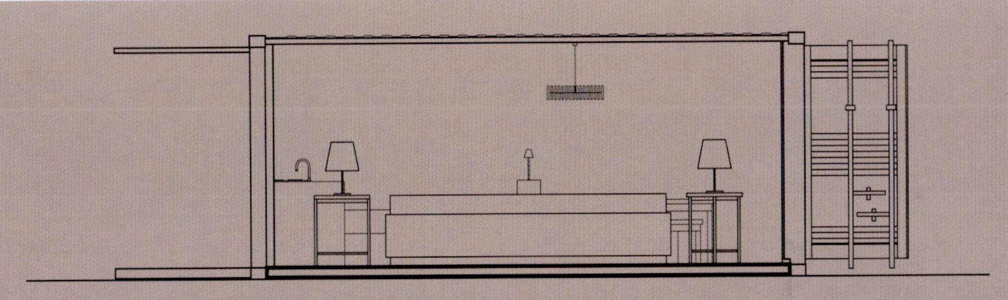

SIDE ELEVATION OPEN

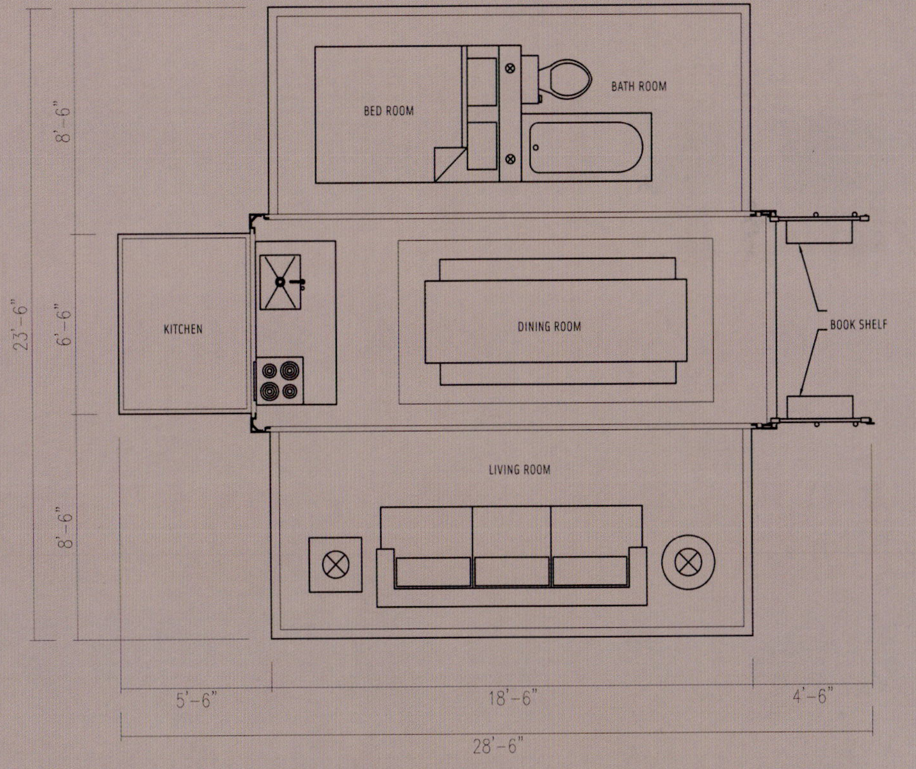

FLOOR PLAN OPEN

1 CONTAINER

NAME➔CONTAINING LIGHT

USE➔MOBILE PRESENTATION UNIT ARCHITECTS➔EER ARCHITECTS, GEERT BUELENS & VEERLE VANDERLINDEN LOCATION➔MOBILE UNIT YEAR➔2005 CLIENT➔KREON

The Belgian lighting manufacturer Kreon has a long tradition of exhibiting its products to potential buyers. Usually, the setting up of the exhibition stand and arranging the products into the desired installation is a time-consuming task. To tackle this very problem, the EER Design team have designed an easy to transport, eye-catching "push the button" pavilion, which is but a 40' ISO container. This mobile showcase container unit can easily be transported from one exhibition site to the next by truck.

When on site, the container is jacked up by four 5m-high (16.4 ft) computer-controlled hydraulic cylinders, which are integrated into the container. The container is lifted from the truck and deposited onto the exhibition floor, where the container base is released from its body. This body is subsequently lifted to its final height of 5m (16.4 ft).

The container is packed with exhibition cabinets, which, when taken out, double the volume of the original container, so that a rather small initial volume is converted into a spectacular architecture. This container-machine has been designed and manufactured in the contemporary tradition, where small media explore great horizons.

The container's exterior is a robust protective shell of welded steel beams and folded plate, with visible mechanisms showing muscle. The interior is the opposite, an exquisite atmosphere of polished stainless steel and white leather.

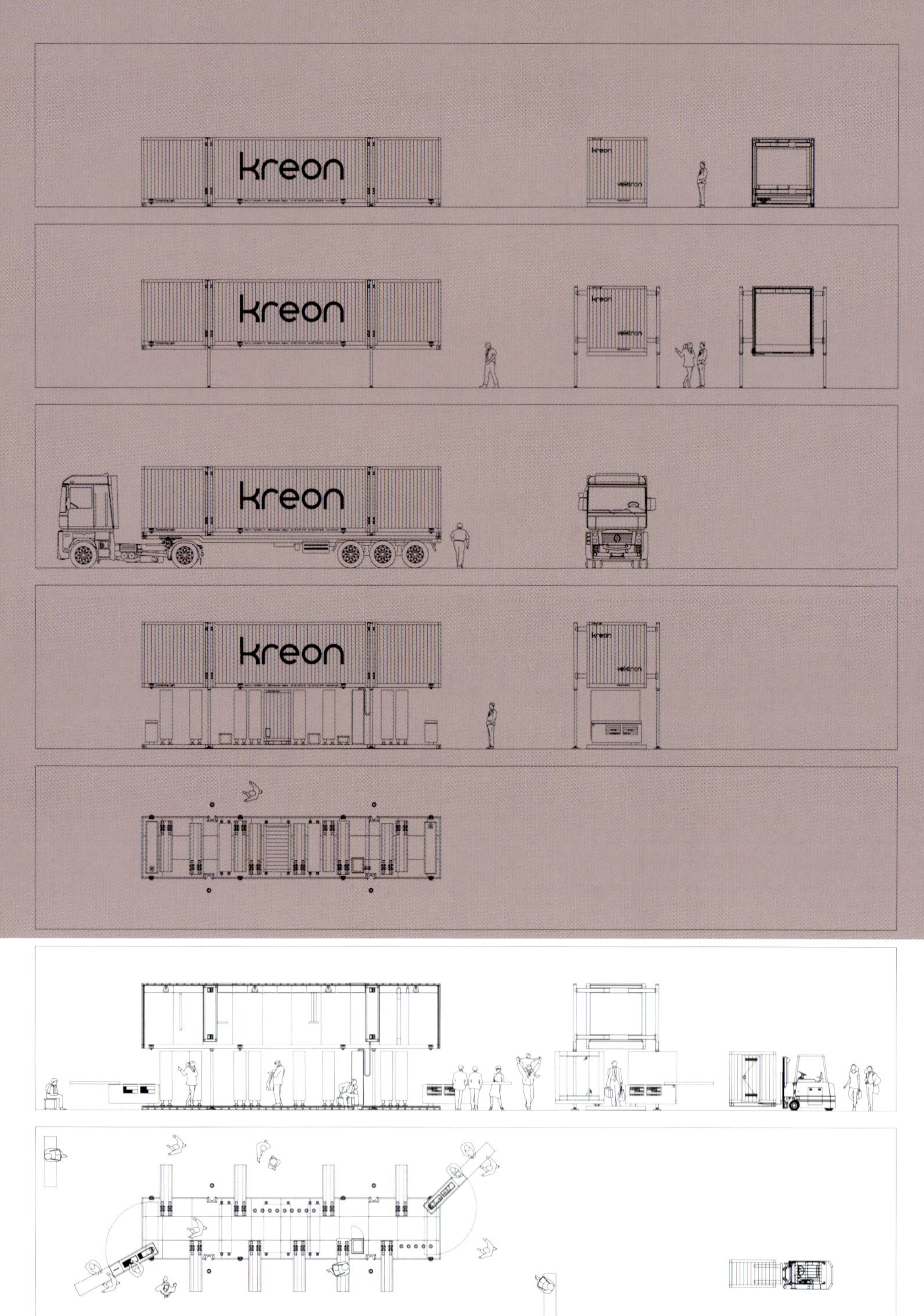
kreon
kreon
kreon
kreon

kreon
KRONE

NAME FHILTEX
USE HOUSING UNIT ARCHITECTS MMW ARCHITECTS LOCATION CAN BE ERECTED ANYWHERE YEAR 1995

The beginning of the 1990s, when this project was set up, saw the shaping of a new generation of people, a generation that grew up in the face of globalization. Distances between continents shortened, borders began disappearing, and free trade resulted in a significant increase in the flow of goods. Most of these goods were transported by means of the building block of globalization – the ISO container, which also serves as the base construction of this pavilion.

Fhiltex is one of the first approximations of a home for the urban nomad – for the person that does not settle in one place for life, is flexible and has limited spatial needs, but cares about the environment. This housing unit is completely self-sufficient, having solar panels supplying it with energy, as well as drinking water and waste-water tanks, both of which can be kept in the interior during transportation. The pavilion is composed of two containers; the larger, 40' one, has a half of the smaller, 20' one, attached to each side. Together with the balconies, the structure comprises 50m² of effective surface area, offering a complete survival kit to an urban nomad: bathroom, kitchen, pantry, etc.

Solid foundations are the sole prerequisite for the setting up of this housing unit. The pavilion is then perched up on slender pillars, its very position high above the ground displaying its independence from any single plot of land. Much like an agile mosquito, it is always prepared to casually detach its slight legs and fly away; and like a true transmitter, this Norwegian mosquito has carried the infectious virus of container architecture all around the world.

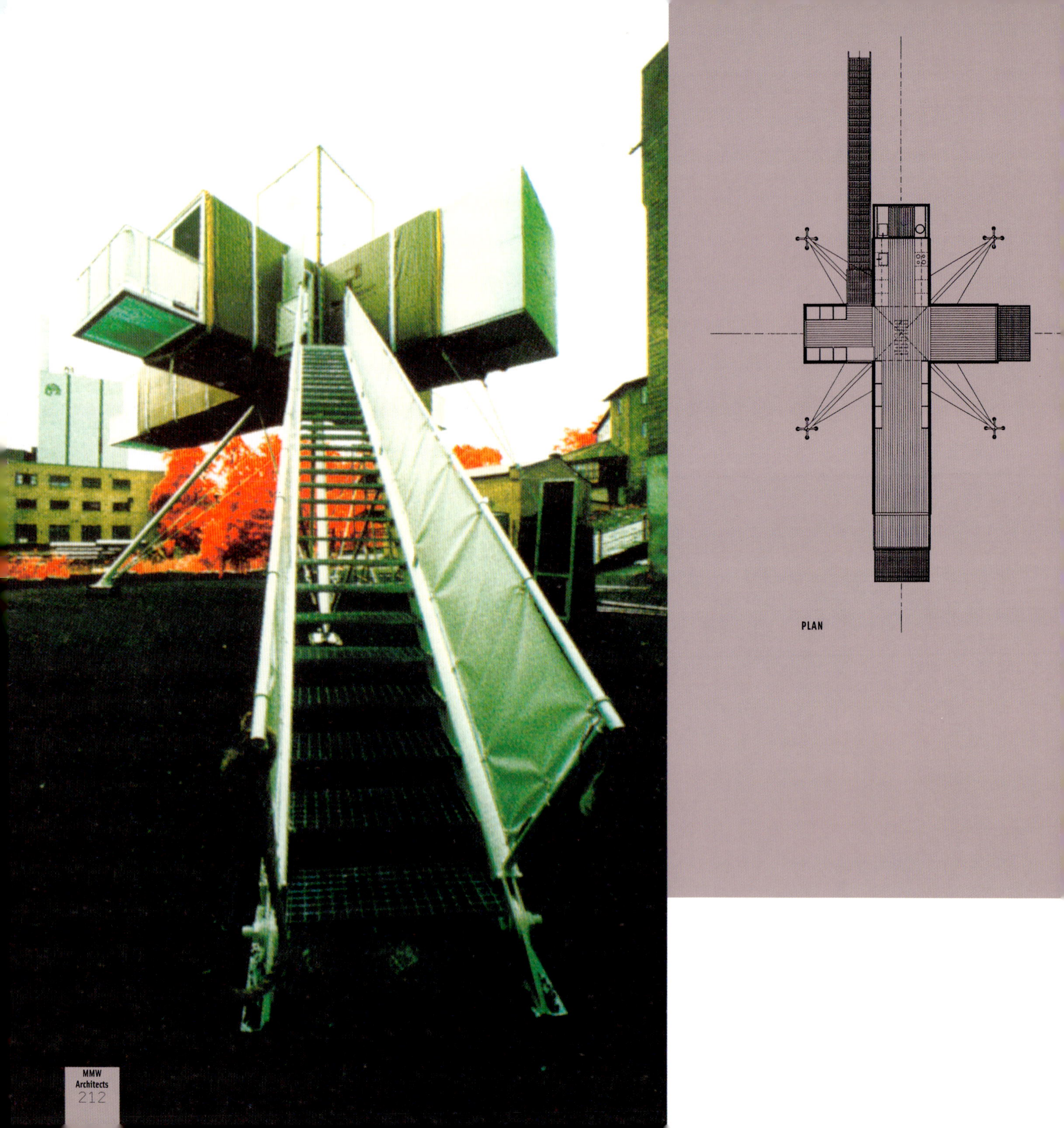
PLAN

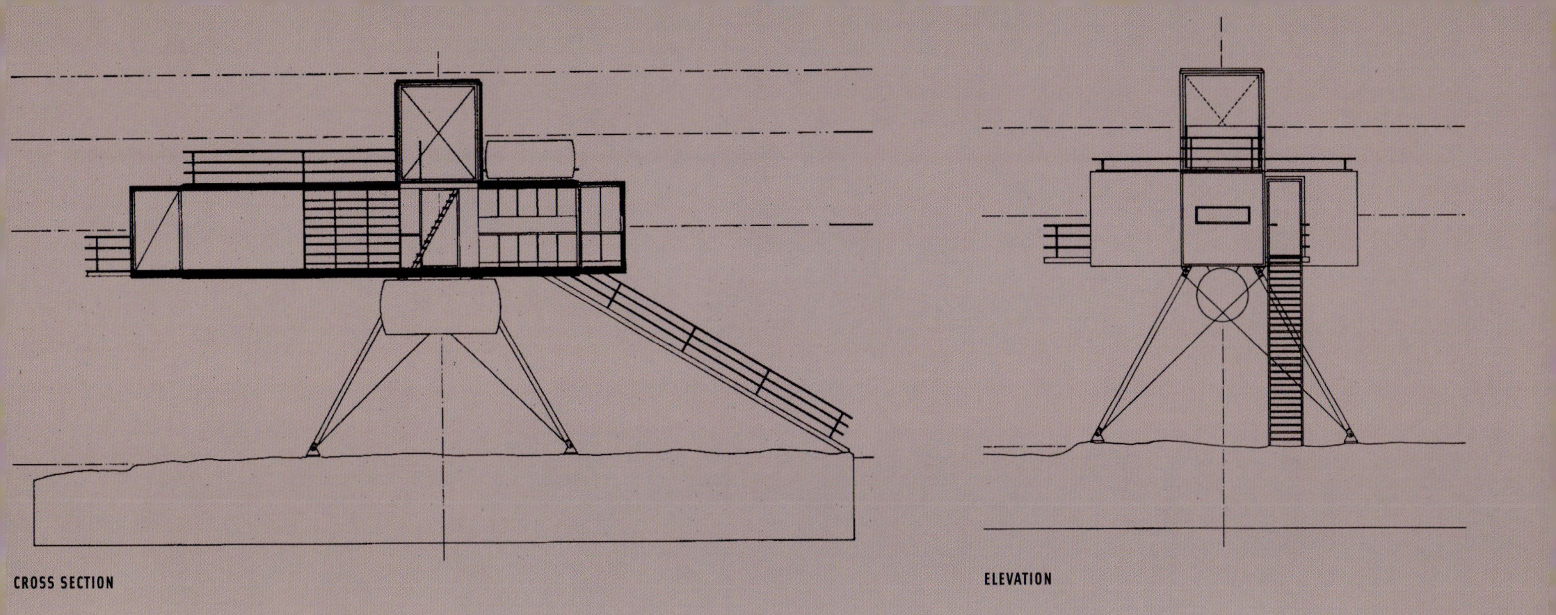
CROSS SECTION
ELEVATION

Helly Hansen
SPESIALPRODUKTER
Helly Hansen
SPESIALPRODUKTER

1 CONTAINER

NAME ➔ **MOBILE DWELLING UNIT**
USE ➔ APARTMENT ARCHITECTS ➔ LOT-EK
LOCATION ➔ EVERYWHERE YEAR ➔ 2002 CLIENT ➔ PRIVATE

Mobile Dwelling Unit (MDU) is the prototype of a relocatable dwelling that, much like a giant briefcase, contains the entire necessary infrastructure enabling a person to live/work in it, as well as pack it up and transport it. It is targeted at people who move a lot, but stay in one place for too long to be in a hotel and for too short a period to settle down permanently.

MDU is a modified 40' shipping container. The façade has openings accommodating several modules, which slide out of the container like drawers and create a spacious interior for when the unit is in use. Each module serves a particular function, enabling the unit's occupants to sleep, work or cook. During transportation, the modules are pushed inside the container and the MDU resembles any other container on a worldwide standardized shipment. But when the container arrives on site, the modules are extracted from the body of the container and the MDU extends into the entire volume of a 40' container, thus becoming a spacious home. The interior of the container and the modules is fabricated entirely out of plywood and plastic coated plywood, including all fixtures and furnishings. Fluorescent lights are sunk flush with the ceiling and floors, while natural light enters the MDU through horizontal windows cut out of individual modules.

The MDU was conceived as comprising a huge system of such housing containers, travelling from city to city and being stacked into special MDU vertical harbors at the final destination. The harbor is a multi-level steel rack, the width of one container and as long as a particular lot permits. The linear container colony is also furnished with vertical distribution corridors that fit staircases, elevators and all other systems, such as water, power, data and sewage. The MDU units are mounted onto their specific location within the harbor by a crane that slides parallel to the rack for the entire length of the linear structure. Steel brackets support and secure MDUs in their assigned positions, where they are plugged-in to connect to all systems. As MDUs come and go, the temporary patterns of this container structure change continuously, reflecting the ever-changing composition of these colonies scattered around the globe.

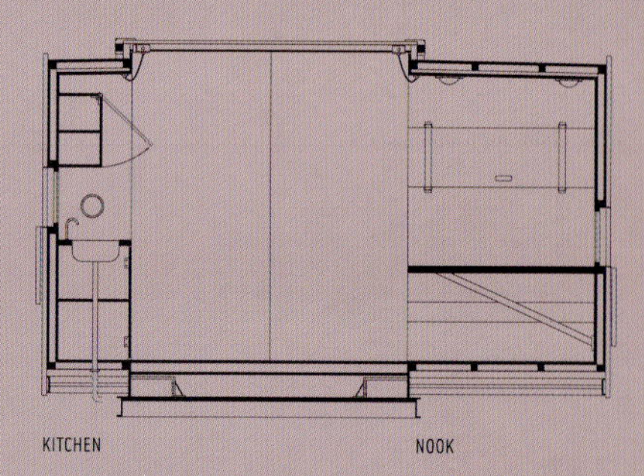

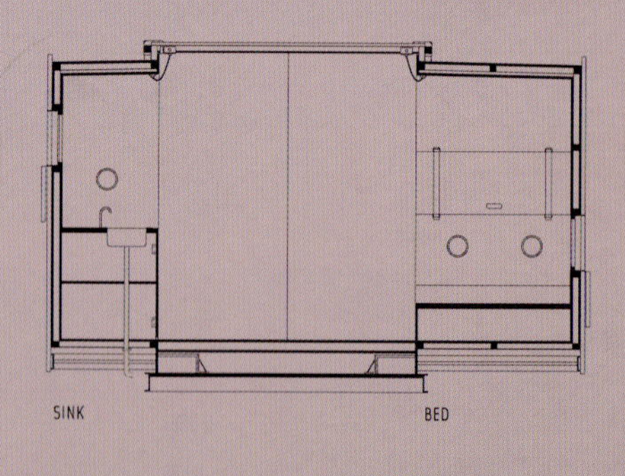

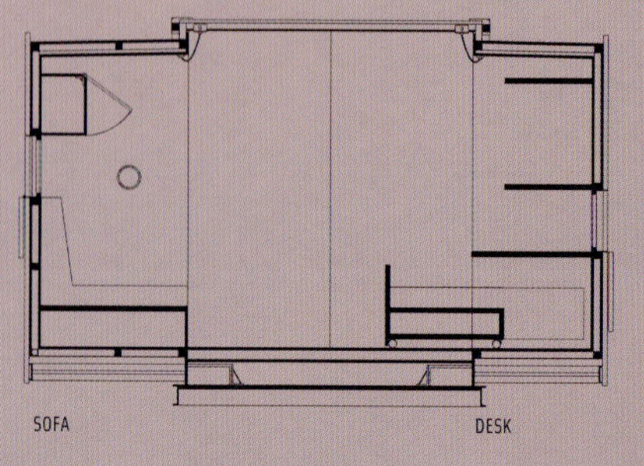

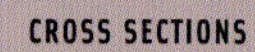

CROSS SECTIONS

MDU VERTICAL HARBOUR

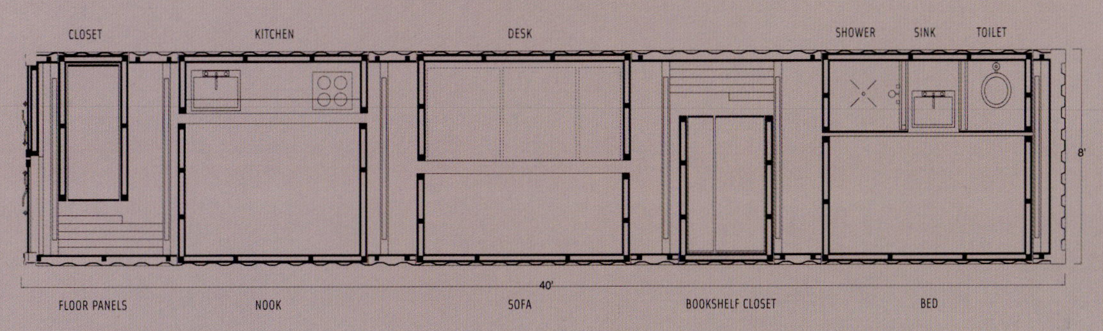

FLOOR PLAN CLOSED

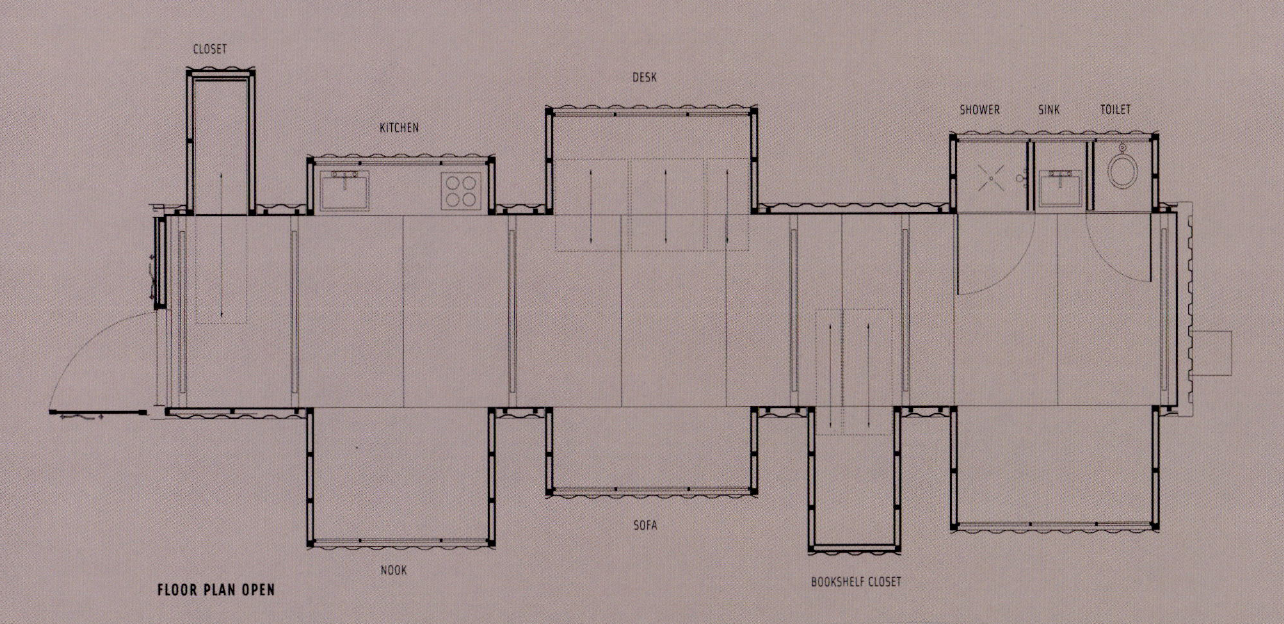

FLOOR PLAN OPEN

1 CONTAINER

NAME PORT-A-BACH
USE BEACH HOUSE ARCHITECTS ATELIER WORKSHOP LOCATION NEW ZEALAND
YEAR 2007

A 'bach' is the name given in New Zealand to small, often very modest holiday homes or beach houses. They represent an iconic part of the country's history and culture, especially of the mid 20th century, when they symbolized the beach holiday lifestyle, which became more accessible to the middle class in that period. In the 1950s, better roads and more available cars allowed families to go on beach holidays. Such holidays became increasingly popular and since people often chose the same beach every year, small beach houses – baches were eventually erected on these spots.

Baches are nearly always small structures, usually made of cheap or recycled material like fibrolite (asbestos sheets), corrugated iron or used timber. They can also be made using a caravan car as the core of the structure, and have extensions built on to that; even old trams were sometimes used. Port-a-Bach, a 20'-ISO-container holiday shed seems to be an appropriate continuation of New Zealand's beach house tradition.

Port-a-Bach features all the advantages of container architecture – it is prefabricated, ready-made and mobile. Its environment-friendly features include low site impact and the fact that it is off-grid, being an energetically self-sufficient unit that uses solar power. Resting inside a compact container shell, it is much sturdier than the older baches. When not used, it looks like an ordinary container on the outside, without giving away any clues as to its contents, and is thus not tempting to potential burglars. One of the container's longer sides opens up to reveal an all-glass façade, and when in the open position this side of the container forms a terrace. The well-illuminated interior is furnished mainly in wood, which creates a pleasant indoor atmosphere. Fold-down furniture is stored inside built-in wardrobes in the walls, which slide open when needed. Port-a-Bach also has an exterior canvas screen system that provides shelter to the deck area, allowing for comfortable indoor/outdoor passages of its occupants.

The container-made bach can house a family of four. And since it boasts so many bach-qualities, this modern prototype will surely receive copies along the beaches of New Zealand.

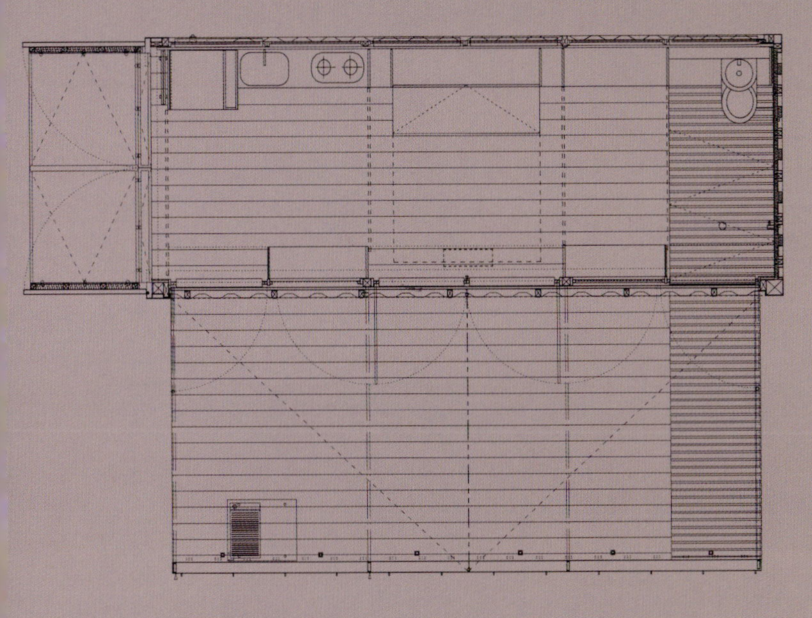

FLOOR PLAN OPEN

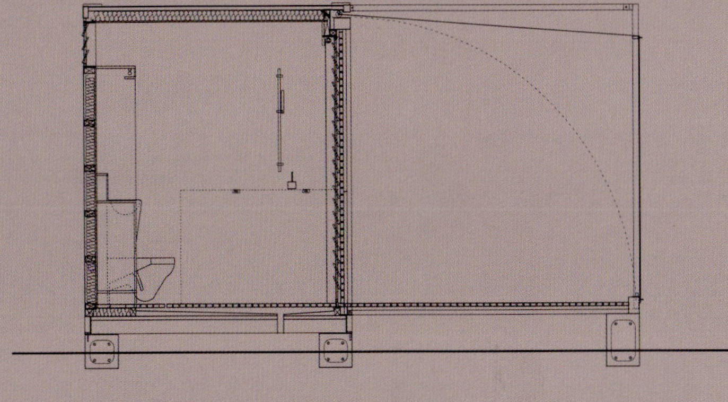

CROSS SECTION OPEN

NAME ALL TERRAIN CABIN
USE CABIN/EXHIBITION PAVILION ARCHITECTS BARK DESIGN COLLECTIVE LOCATION VARIOUS YEAR 2006
CLIENT BARK DESIGN COLLECTIVE

BARK is a non-profit collective of designers and businesses committed to raising the profile of Canada and Canadian design internationally. To make the country known for more than beautiful mountains, maple syrup and raw logs, they decided to showcase Canada's innovative design talent and send it on world tour. An ISO-container cabin, the ATC, was chosen for this purpose and sent to tour international designer shows and environmental conferences, urban centers as well as smaller country towns, for the period of five years.

The All Terrain Cabin (ATC) fits into any surroundings, with or without municipal services, being a self-contained and environmentally-friendly product. It is a cabin teaching us how to leave a lighter footprint on the planet, having a collector for rainwater, a facility for reuse of grey water, composting toilets, solar panels and a generator running on biodiesel. This off-grid container home is built of aluminum, wood and glass, and is easy to transport. During transport, it looks just like a standard shipping container, but once it arrives on site, the walls unfold to reveal a comfortable mobile home suitable for a family of four and a pet. It comprises a terrace, an ISO container and a tent. The container is the technological nucleus of the unit, holding all installations, the toilet and kitchen, while the living area and bedroom are in the tent.

The ATC is a genuine example of Canadian manufacture, being furnished solely by the products of 56 Canadian designers and companies, which contributed everything from bathroom ceramics, furniture and lights to a Canadian-made dustpan, bicycle, thermos, etc. The ATC stands for a 21st century concept of living and introduces the world to a broader and more accurate display of Canadian culture, thereby creating new economic opportunities for Canadian individuals and businesses of all sizes.

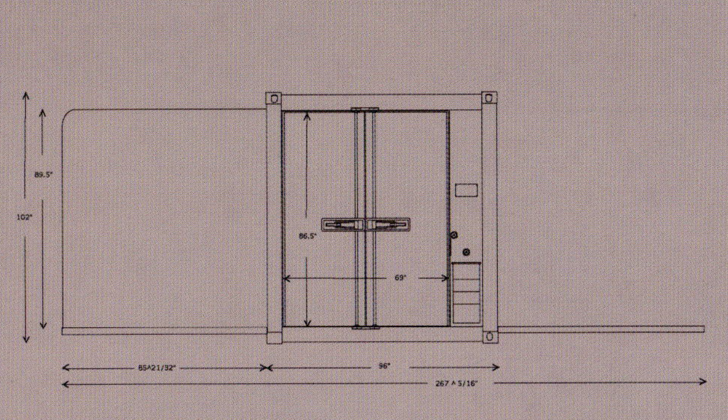

LEFT SIDE OPEN

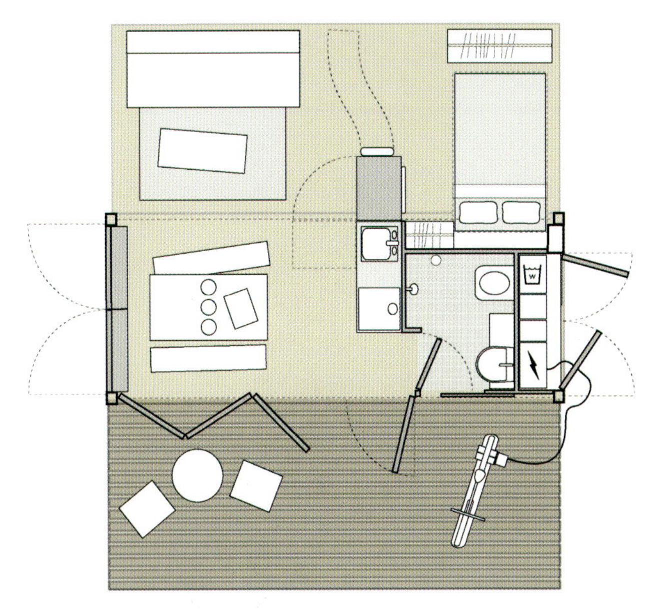

FLOOR PLAN OPEN

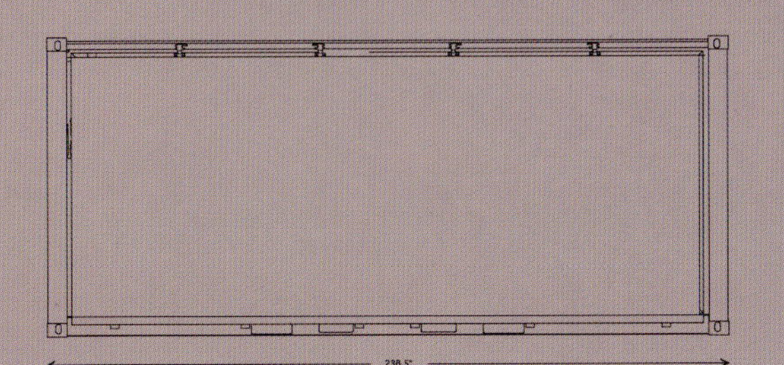

BACK VIEW OPEN

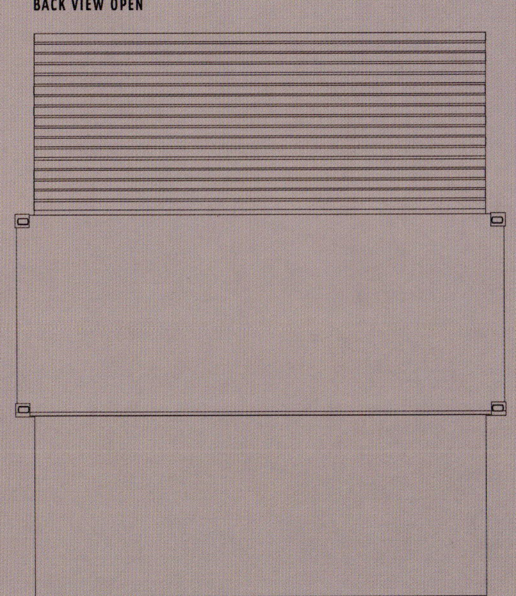

TOP VIEW OPEN

NAME➔**CONHOUSE 2+**
USE➔WORK/LIVE UNIT ARCHITECTS➔JURE KOTNIK
LOCATION➔SLOVENIA YEAR➔2007
CLIENT➔TRIMO D.D.

Sky high prices of real estate in the contemporary world have stimulated the search for and development of alternative housing solutions. One such attempt is the ConHouse system of small-size housing container units, which takes the housing/office ISO container to the next evolutionary level. As opposed to the other container projects, which mostly feed on the excess of available cargo containers, ConHouse pushes the development of containers manufactured especially for housing and office purposes. Even though the past 20 years have significantly changed the face of architecture, the basic form of these containers has remained the same. The ConHouse system now upgrades them to enable a quality of living comparable to classical housing. This is achieved mainly through rationally designed ground plans, carefully selected materials, a well-lit interior and the customized outer appearance of individual units. By enabling its occupants to co-shape their compound unit according to their particular needs, ConHouse takes after the inventive automobile industry, where car owners can chose the components they desire, and after the IKEA-developed winning recipe of interior design.

The 2+ pavilion has been intentionally designed to stand out with its contrasting colored criss-cross exterior, to promote from afar this type of construction. The 2+ is a two-level mini housing/office unit composed of two containers perpendicular to each other. 2+ thus shows that a minimal number of containers combined in an innovative fashion offers fresh yet functional architectural solutions. The upper container provides a projecting roof above the entrance as well as serves to shelter the back terrace. The ceiling of the bottom container is also a terrace of the first floor. The pink-dotted façade illustrates the wide range of possibilities for tailor-made exteriors, the choice of which is as simple as deciding about which mobile phone cover to put on. The system's modular nature enables containers to be added to or subtracted from the compound as needed, so that the ConHouse can grow or contract depending on the actual spatial needs of the people using it. Lower prices of such live/work units make them competitive as compared to traditional housing and are intended to increase the number of home-owners, who can then use the extra cash to expand their living space or invest more into interior design.

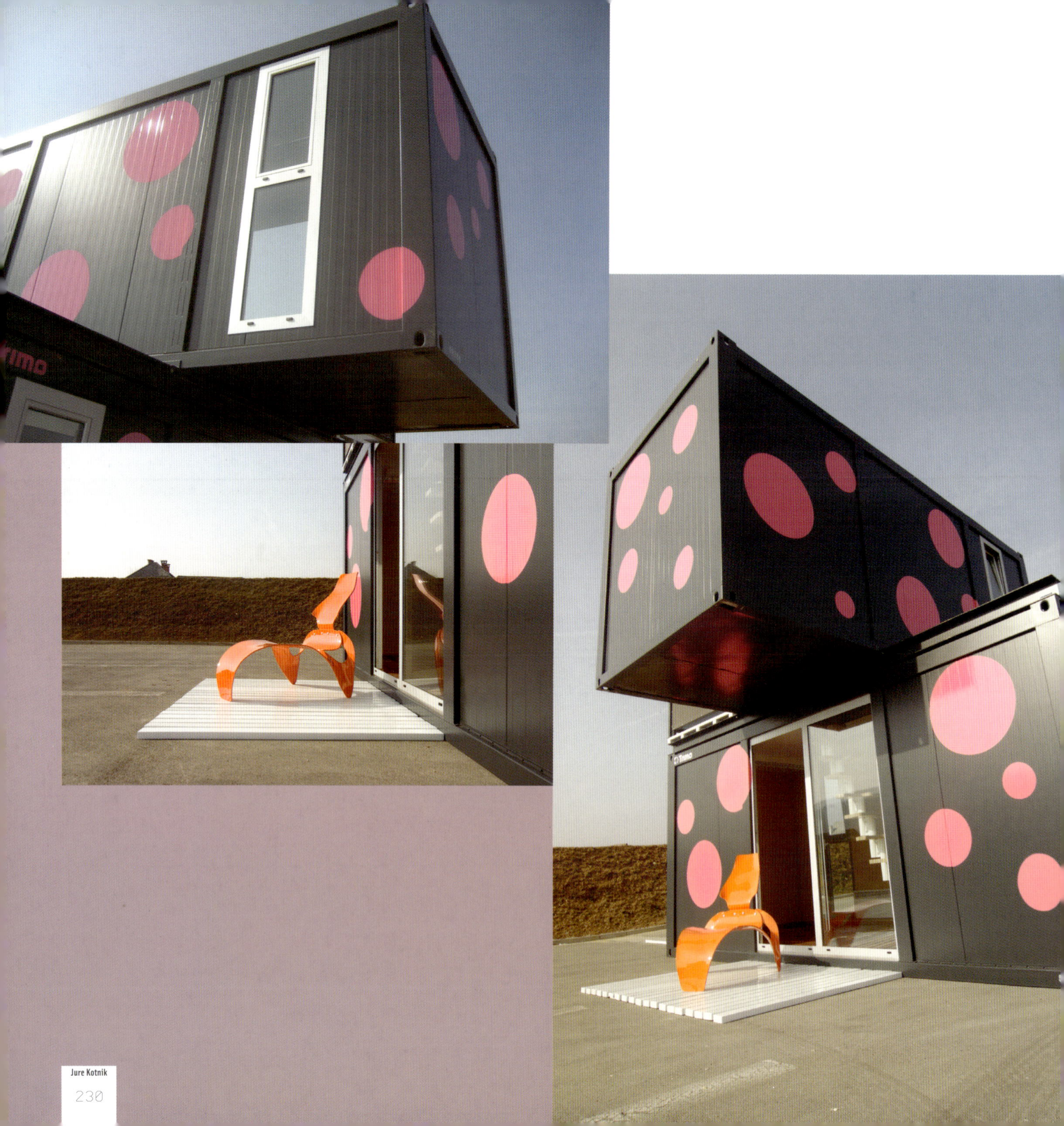
Trimo

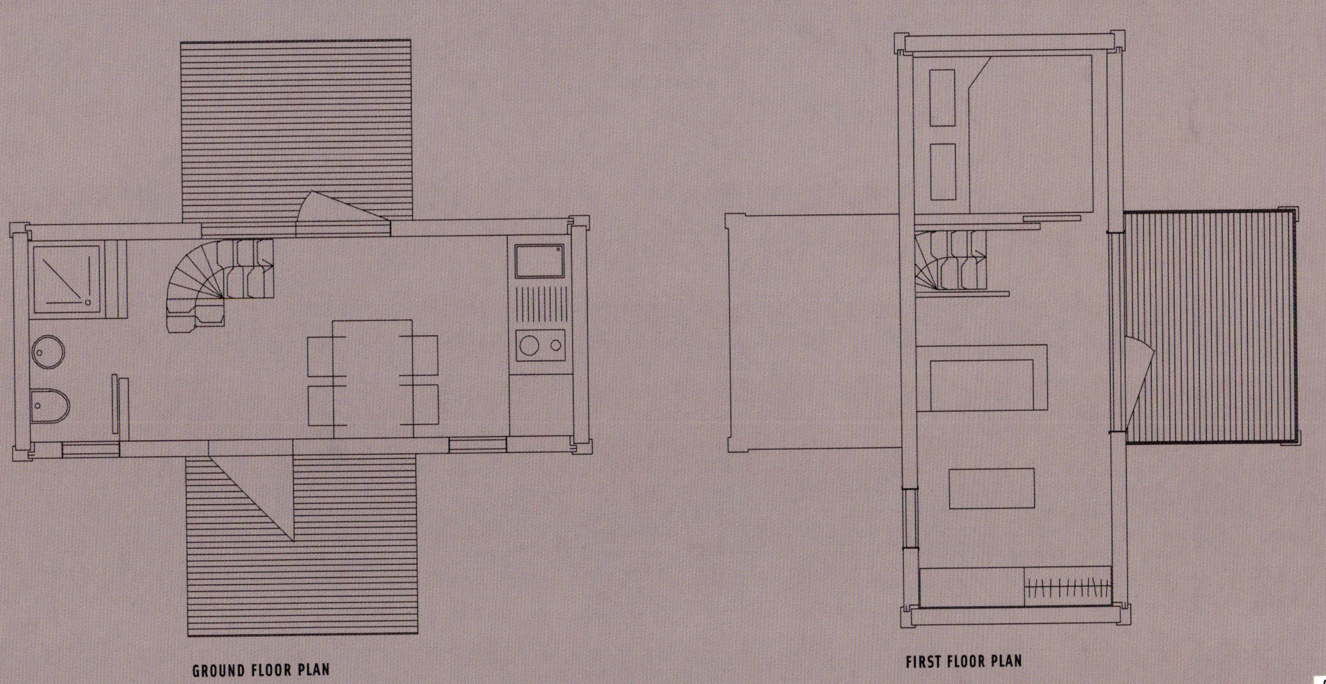
GROUND FLOOR PLAN
FIRST FLOOR PLAN

If entire apartments can fit into a container, then so can individual parts of an apartment. Such parts may then be taken away from or reunited with the whole, or stand alone permanently, like this container sauna by the Castor Canadensis trio. The Canadian product brings to life in a single container the predictions made by Jones and Partners architects on container modules that would accommodate parts of a flat and facilitate such functions as cooking, personal hygiene, entertainment and last but not least, having a sauna.

The Saunabox is a simple enough combination of a 10' ISO container and the traditional wooden sauna. The container's sturdy outer shell is constructed of Cor-Ten steel, well known for its sculptural properties and stable rust finish, which makes the Box weatherproof without the need for any additional protective layering. From the outside, the Box looks like any other shipping container, except for a small window and chimney.

When in use, standard container doors are pulled open to reveal a tiny entryway with the shower and a wood-fired stove. Inside, the sauna is spacious. The interior, including the bench, is cedar wood and illuminated through a window and glazed doors. Each Box is site specific and custom-made. To underline its customized nature, Castor Canadenis (named after a species of beaver) have created a Saunabox with a plug-in for an electric guitar and with built-in speakers playing music from an iPod, while offering additional handmade items to furnish it with, such as a beaver-whittled stone stool, stone sink, wood and metal work.

The Saunabox is nature-friendly and suitable for use off-grid. It is powered by solar panels on the roof, where rainwater tanks can also be mounted to supply shower water. It can easily be transported to any location, needing very little site preparation.

1 CONTAINER

NAME→**SAUNABOX**
USE→SAUNA ARCHITECTS→CASTOR
LOCATION→ONTARIO, CANADA
YEAR→2007 CLIENT→VARIOUS

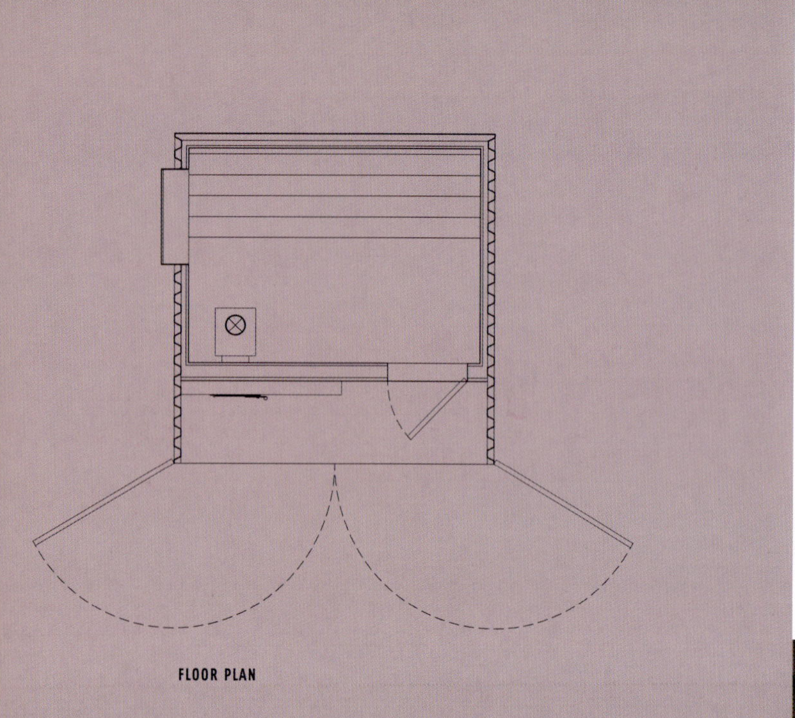

FLOOR PLAN

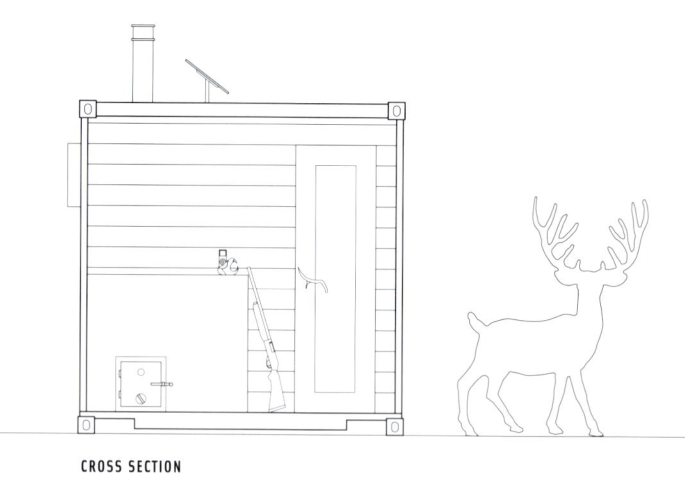

CROSS SECTION

NAME HOORN BRIDGE

USE PEDESTRIAN/BICYCLE BRIDGE ARCHITECTS LUC DELEU – T.O.P. OFFICE LOCATION HOORN, THE NETHERLANDS YEAR 1990
CLIENT STICHTING DE ACHTERSTRAAT, HOORN

Artist and architect Luc Deleu is renowned for container sculptures installed into public space, which he takes as his natural work environment. These installations are conceived as temporary, since they aim to break down architecture's static character and to express his concern for the ongoing shrinking of open space.

His container constructions comprise architectonic archetypes, such as an arch of triumph, gateway, obelisk, bridge, etc, that traditionally make up the urban environment. His constructing these structures monumental significance from containers creates an ironic tension between the concept of the archetype and the material it is constructed from.

The Bridge at Hoorn, probably the single one of its kind in the world, was an installation for the exhibition "For Real Now" in Hoorn, The Netherlands (1990). It was constructed as a pedestrian and bicycle bridge for the time of the exhibition, and set up on the spot where a bridge was of obvious need to the inhabitants. The riverbed was bridged by two interconnected red ISO containers, clamped between the riverbanks in mid air. The containers had to be placed at the correct angle to each other, and a load-distributing seating was also needed to mobilize a large part of the embankment. This helped allay the fears of the municipal authorities, who were concerned whether the construction could defy the force thrust on it by people walking or cycling through it.

NAME CONTAINER HOUSING
USE APARTMENT ARCHITECTS GUSTAU GILI GALFETTI LOCATION BARCELONA, SPAIN
YEAR 2005 CLIENT CONSTRUMAT

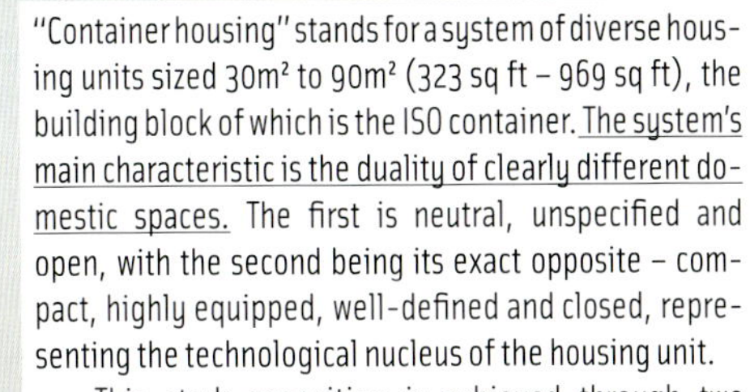

"Container housing" stands for a system of diverse housing units sized $30m^2$ to $90m^2$ (323 sq ft – 969 sq ft), the building block of which is the ISO container. The system's main characteristic is the duality of clearly different domestic spaces. The first is neutral, unspecified and open, with the second being its exact opposite – compact, highly equipped, well-defined and closed, representing the technological nucleus of the housing unit.

This stark opposition is achieved through two clearly differentiated construction systems. The first includes the cost-efficient modular structure of pillars and forged reinforced concrete. The second consists of a compact core composed of recycled 20' ISO containers, where due advantage is taken of all the merits of these ready-made boxes (modular, portable, rugged) in transposing them into the domestic sphere.

The present prototype, sized $30m^2$, was set up in 2005 at Barcelona's Construmat building exhibition. It consists of an orange 20' ISO container, and a large concrete block with two shorter and two longer sides, one side of which is all glass. The container pierces through the centre of the structure, thus separating the remaining space into halves – they are connected by the hall inside the container, which also serves as the entrance. The container further contains a wardrobe, bathroom with toilet, and kitchen, while the purpose of rooms outside the container remains to be designated arbitrarily.

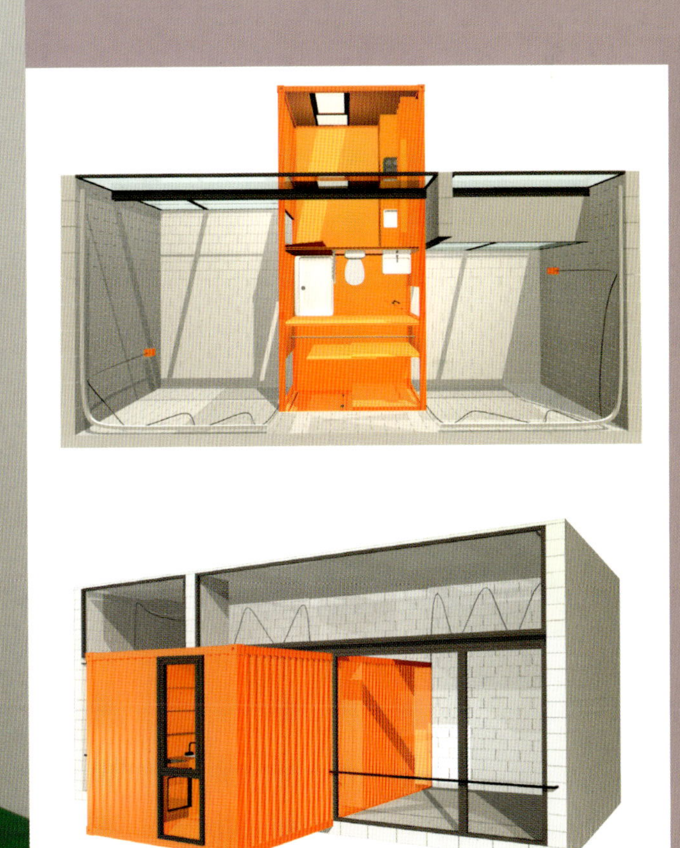

gustav gili galfetti
CONTAINER HOUSING. 30 m2
RQ
42731
JC1
N4

VTSX 241

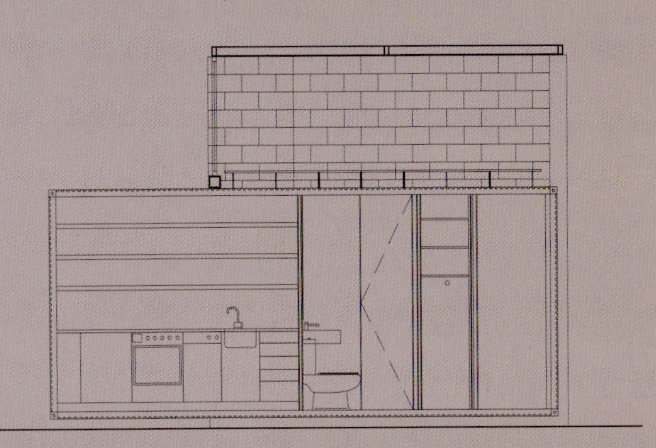

LONGITUDINAL SECTION

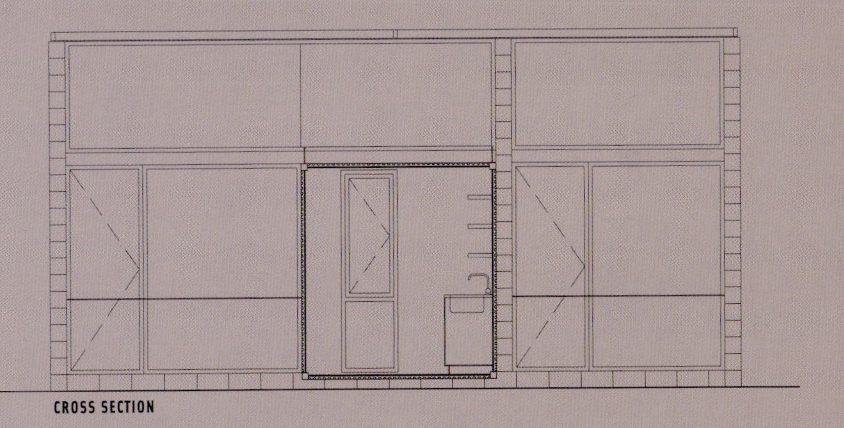

CROSS SECTION

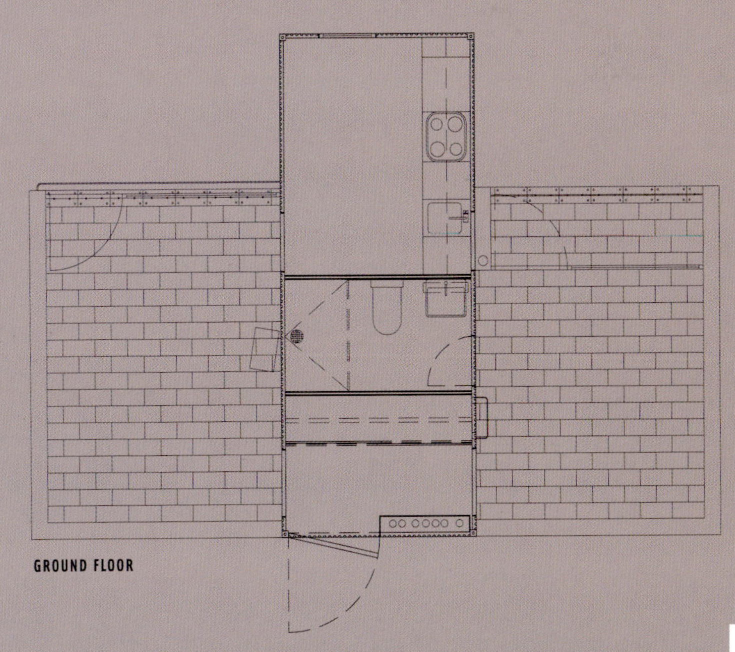

GROUND FLOOR

NAME: **HOME SWEET HOME**
USE: EXPOSITION PAVILION ARCHITECTS: COMA, GILI, O'FLYNN, LELYVELD, SCHULZ-DORNBURG LOCATION: BARCELONA, SPAIN
YEAR: 1996 CLIENT: VIA/FAD

The "Home Sweet Home" exposition pavilion, assembled from seven orange containers in apparent geometrical disorder, stood on a square in front of a cathedral in Barcelona. Containers, otherwise typical of construction sites, contrasted vividly with the monumental buildings in their immediate vicinity. The pavilion temporarily transformed public space into an exposition. Its aggressive color and vivid composition attracted the looks of passers-by, inviting them to step inside.

Erected in 1996 alongside the XIX Congress of the International Union of Architects, it housed an exhibition that critically highlighted how neglected housing had become as the basic (minimum) element of any urban environment (urban aggregation). The "Home Sweet Home" containers were connected in such a way so as to lead the visitor through the entire exhibition area and past all the key players in the construction scheme; each container presented one element from the construction chain – from developers to architects, from designers to the media, and finally to the end-user: the dweller.

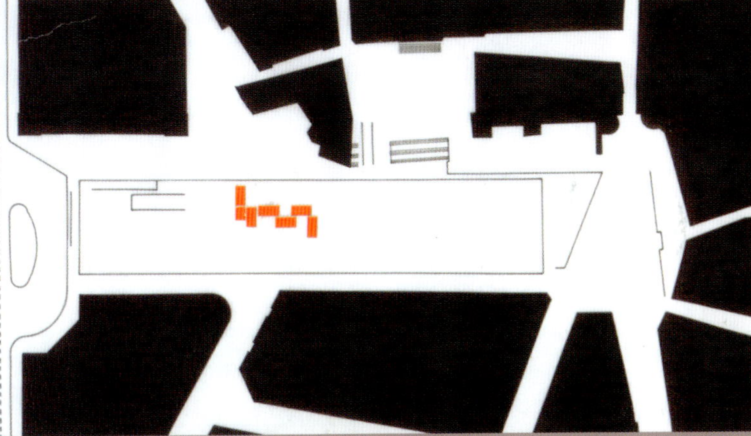

Located on the corner of New York's 5th Avenue, which symbolizes wealth and prosperity, and one of the city's major cross town streets, the 42nd Street, is a would-be vertical container mall. In a spot where you expect classy construction materials such as marble or glass, the installation of containers – the linking elements of the product & transport chain, which an end user is not at all interested in, let alone a 5th Avenue frequenter – seems a contradiction in terms. This is one bold idea…

… and, if realized, it would be an intervention that could not go unnoticed, if for its sheer size. A vertical mall takes on the form of a giant red screen composed of 125 containers. Nine levels of containers are stacked along the facade of a classical Manhattan 1930s building, but without touching it directly; a system of catwalks, stairs, and elevators is wedged between the container stack and the wall of the building to make up the circulation and a series of outdoor public spaces. The containers are placed in an undulating manner, and even missing in some parts, thus allowing the exchange of air, light and street views, as well as animating the red facade.

The main advantage of such vertical malls is their capacity to conjoin with existing buildings and their small ground plan. LOT-EK therefore suggest that such ivy malls are perfect for left-over empty lots throughout a city.

128
CONTAINERS

NAME➔ **CONTAINER MALL**
USE➔ **RETAIL** ARCHITECTS➔ **LOT-EK**
LOCATION➔ **NEW YORK, NJ, USA** YEAR➔ **2003**

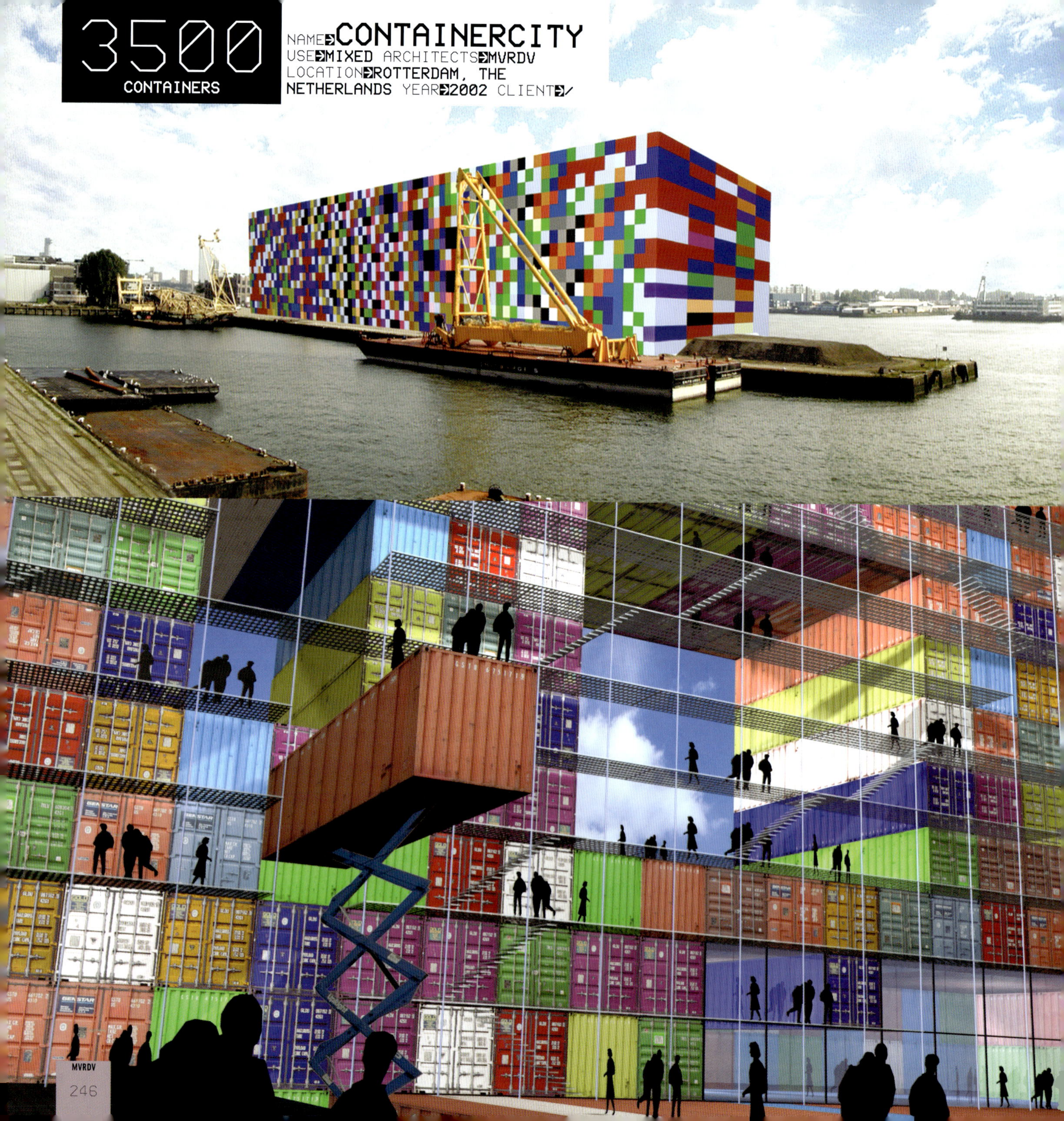

3500 CONTAINERS

NAME→CONTAINERCITY USE→MIXED ARCHITECTS→MVRDV LOCATION→ROTTERDAM, THE NETHERLANDS YEAR→2002 CLIENT→/

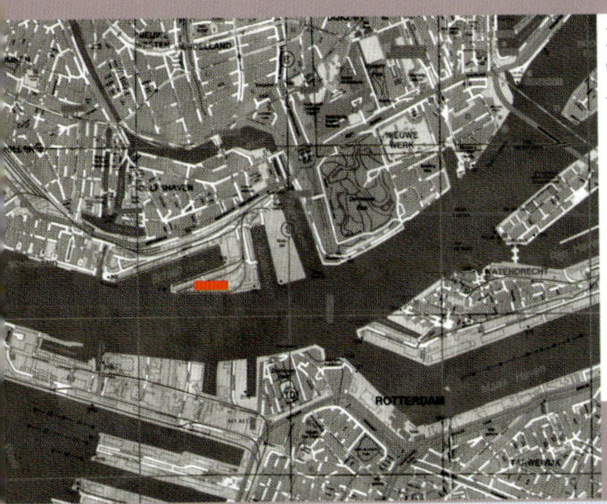

3500 containers, temporarily subtracted from the world-wide flow of trade, have been gathered in Rotterdam to form a megacontainer of the scale of a city: Container City. Containers serve as floors, walls and ceilings of an immense structure of unusual proportions. They are connected by cables, which create sufficient tension to maintain 6-container-long hollow 'beams' that span the hall, while also allowing for the stacking of 15 units on top of each other.

The visitor is thus surrounded by thousands of units, each representing a direct link between the intimate and the grand scale of the hall. This giant "beehive" hosts 3500 niches for sleeping, eating, exhibiting, performing, and contains enough room for hotels, bars, galleries, a spa, conference halls, shops, business units, ateliers, schools, crèches. Each container unit can be accessed through galleries, "construction-elevators" or staircases. Parts of the megastructure can easily be removed by placing them on rails and hoisting them away, thus creating giant "windows" into the surrounding.

GOLD
GENSTAR
MAX GR.
TARE
NET
CU.CAP.

GLDU
097152
42G1
MAX.GROSS
TARE
PAYLOAD
CUBIC CAPA
GOLD

Home Brewery

14 CONTAINERS

NAME PRO/CON PACKAGE HOME TOWER

USE RESIDENCES ARCHITECTS JONES & PARTNERS ARCHITECTS
LOCATION URBAN/SUBURBAN AREAS YEAR 2000 CLIENT HAMMER MUSEUM/UCLA

PROgram CONtainers is the name of a well-defined system of container housing, which recognizes the container's intrinsic qualities such as its mobility, flexibility, sturdiness, prefabrication, modularity, etc. On account of these features, the container is used as the building block for constructions of various shapes and sizes, and it can also adapt well to the relevant building sites. In planning a building, the container's tectonic integrity is preserved, and more attention is paid to the space between containers. Several applications evolved from PRO/CON, including the Package Home series.

Package Home projects decline the idea of common indoor spaces, such as the living room. It is presupposed that in the Internet era residents are individuals who can communicate and socialize online, but who chiefly like to escape into their own worlds. Such independent worlds are therefore accommodated in program-specific containers, depending on the desired activity and manufacturer, whose logo is printed on the outer shell of the building. This results in such units as e.g. a container IBM computing unit, a Sears bedroom, a Disney entertainment unit, a Mattel Barbie unit for children, etc. The desired unit is ordered online, delivered to the chosen location and assembled to the homeowner's specification.

There are three Package Home prototypes: Tower, Ranch House and Stack Shack. The Tower comprises 14 containers stacked into seven levels, and is intended to accommodate students. The entire composition sits on a slewing ring, which makes the Tower rotate as needed. This undermines the myth of architecture as being rooted in spatial and temporal fixedness. In the center of the ring, a vertical void space runs the entire height of the building and is circled by the stairs providing access to individual container units. There are no fixed floors here, but they can be introduced as needed.

The Tower is a realization of Le Corbusier's "The house is a machine for living in", adapted to the present day society through modern technology. The result is a high-tech machine-like building, composed per partes of containers chosen by individual inhabitants.

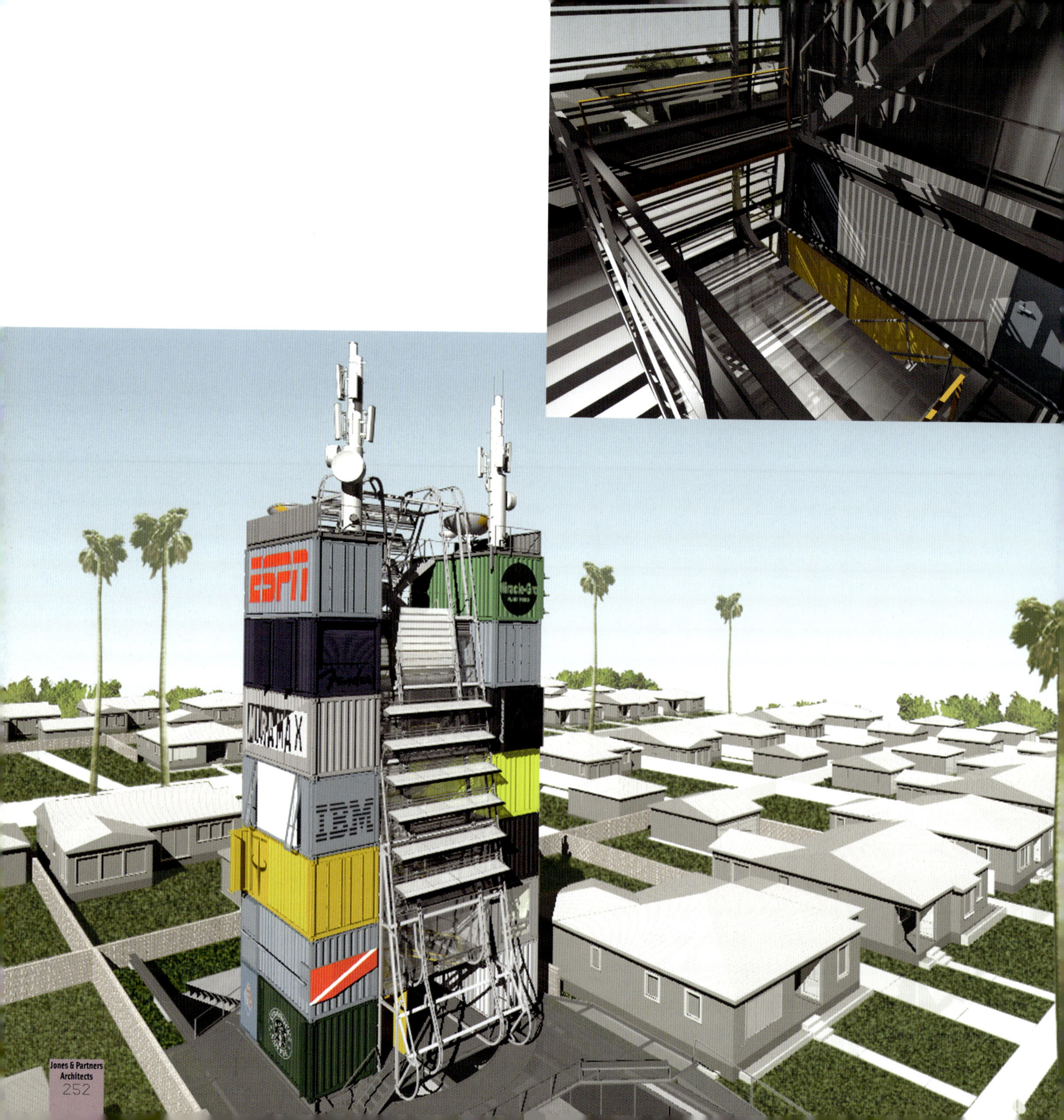
ESPN
MIRAMAX
IBM

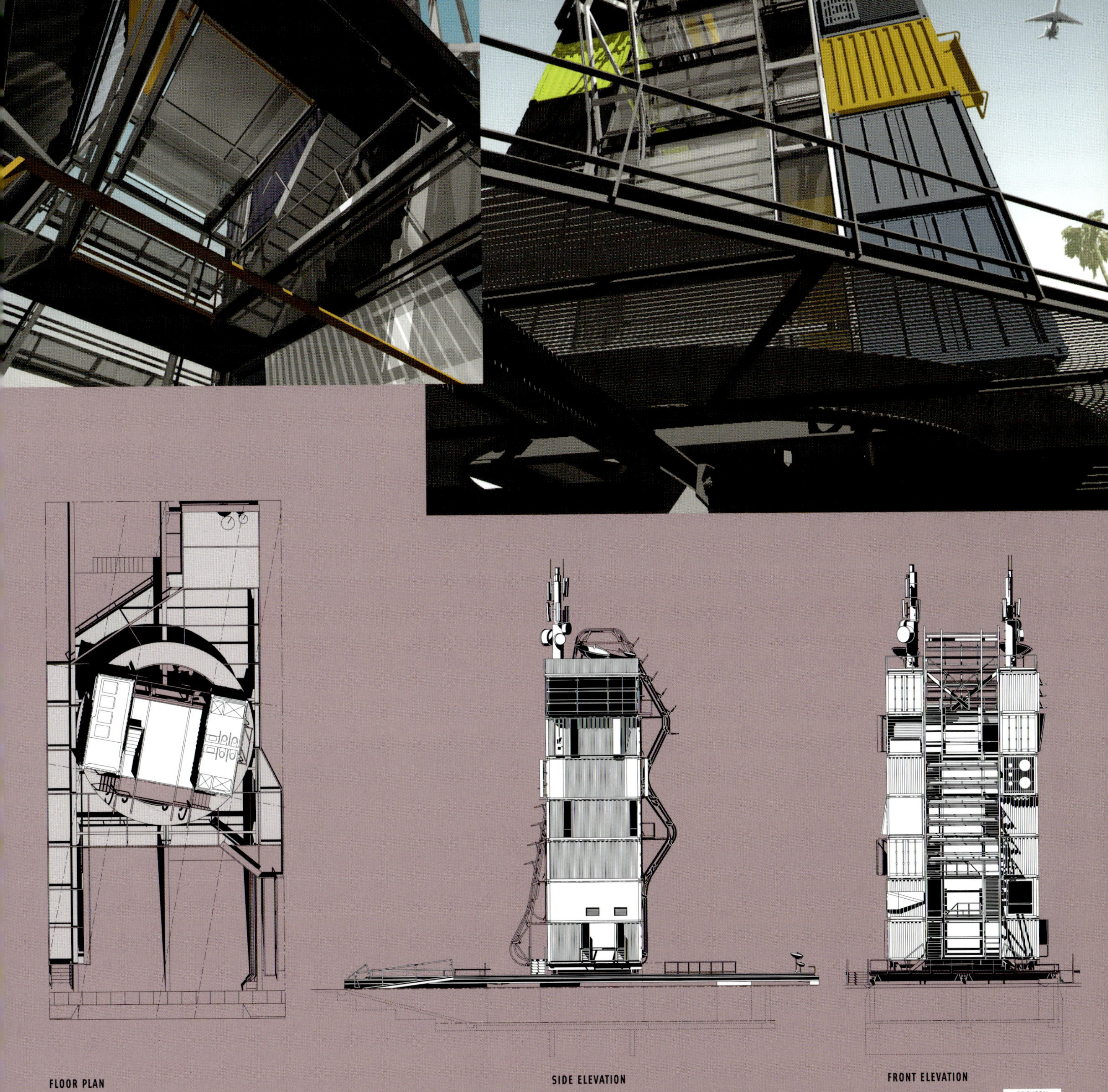
FLOOR PLAN
SIDE ELEVATION
FRONT ELEVATION

BOZU
140
636
8

FIRMITAS!
Utilitas!
Venustas!